T0360558

QUANTITATIVE ECONOMICS IN CHINA

A Thirty-Year Review

QUANTITATIVE ECONOMICS IN CHINA

A Thirty-Year Review

Editors

Shouyi Zhang
Chinese Academy of Social Sciences, China

Tongsan Wang
Chinese Academy of Social Sciences, China

Xinquan Ge
Beijing Information Science and Technology University, China

*The translation of the book has received financial support
from the Innovation Project of the Chinese Academy of Social Sciences.*

World Scientific

NEW JERSEY · LONDON · SINGAPORE · BEIJING · SHANGHAI · HONG KONG · TAIPEI · CHENNAI · TOKYO

Published by

World Scientific Publishing Co. Pte. Ltd.

5 Toh Tuck Link, Singapore 596224

USA office: 27 Warren Street, Suite 401-402, Hackensack, NJ 07601

UK office: 57 Shelton Street, Covent Garden, London WC2H 9HE

Library of Congress Control Number: 2015948928

British Library Cataloguing-in-Publication Data
A catalogue record for this book is available from the British Library.

中国数量经济学30年
Originally published in Chinese by Social Sciences Academic Press
Copyright © Social Sciences Academic Press, 2014

QUANTITATIVE ECONOMICS IN CHINA
A Thirty-Year Review

ISBN 978-981-4675-67-3

In-house Editors: Herbert Moses/Tan Rok Ting

Typeset by Stallion Press
Email: enquiries@stallionpress.com

Printed in Singapore

series of laws including turnpike theorem, fixed point theorem, economic core theory (core, kernel, etc.) and economic chain theory. However, one may wonder whether these laws are valid in the case of imbalance theory. And if they are, do they vary to some extent?

Von Neumann proved turnpike theorem on the premise of constant structure and showed that assets of all sectors increase at the same speed every year. This model has no practical value. In 1985, we adopted a Variable Structure Model under which the economic operation track is changed from a line to a curve. Does turnpike theorem still exist in this case? This question is still to be answered.

Belushi states that "Non-Walrasian Equilibrium is different from anti-Walrasian Equilibrium". Instead, it refers to the fruitful methods applied to the Walrasian Equilibrium under the general assumptions. We should deny general Walras Equilibrium completely and establish an anti-Walrasian disequilibrium Economics. Janos Kornai believes that the private economy is a hard budget constraint which manifests itself as demand insufficiency and supply surplus. Comparatively, the public economy is a soft budget constraint under which income distribution favors individuals and demand surplus and supply insufficiency occur easily. In both cases, there is a gap between demand and supply.

From what we have discussed above, we can elicit two major questions. One is, by means of ownership reform, can we possibly establish a new imbalanced economy that "combines soft and hard budget constraints and bridge the gap between supply and demand" so as to produce good economic results? The other is, since the private economy emphasizes efficiency whereas the public economy stresses fairness, can we possibly achieve a relatively reasonable balance between efficiency and fairness by reforming the ownership system as well as fiscal and taxation policies?

(4) Economic control theory

Studies in this field seem to be sluggish recently. In fact, the expansion of macro-political instruments, the organic combination of various polices, the setting of orientation, focus, strength and tempo of each policy as well as the role of macro-control in the meso-economy and micro-economy are all major issues to be studied from the perspective of economic control theory.

(5) One theoretical hypothesis

After a long period of thought, I proposed an hypothesis of theoretical economics in 1992. According to this hypothesis, game theory, disequilibrium theory, nonlinearity theory, non-stability theory and theory of business cycles seem to study absolutely different things. But actually, they are of a causative relationship with game theory as the cause, disequilibrium theory as the result and nonlinearity theory, non-stability theory and theory of business cycles are the three manifestations of disequilibrium theory. The complicated relationship among them is what modern economics including quantitative economics needs to study.

2. Quantitative analysis

(1) Use input–output analysis technology to study the cyclic economy

The Bureau of Statistics of Hebei Province made some efforts in this respect by reflecting the cyclic economy with one input–output table. But I propose to use multiple tables to reflect the cyclic economy. It is necessary to do pilot work in a large chemistry industrial enterprise and introduce the input–output analysis technology to other enterprises, departments, regions and finally throughout the country.

(2) Conduct microeconomic predictions

Since prices of such goods as petrol, raw material and stocks fluctuate violently, microeconomic predictions are fairly important and difficult. Scholars abroad are doing related research and we should fill the research vacuum in China as soon as possible.

(3) Study disequilibrium measurement

Currently, disequilibrium measurement models in three markets are relatively well-developed. Our question is: How many markets do we have under the market economy system? As the number of markets, the overflow effects between them are becoming complicated and we come across greater difficulties in using the current models. In this circumstance, we need advanced software.

(4) Study nonlinearity measurement

Linearity measurement is only an approximation of the quantitative relationship under the economic system whereas nonlinearity measurement is what reflects the nature of the quantitative relationship. Therefore, we need to study the basic functional relationship and develop a simultaneous system of equation model on the basis of the employment of Single Equation Model.

3. Simulation analysis

Economics is not to become harder and harder, but to become softer and softer. When studying decision-making problems, we should pay more attention to psychological, neurosis and genetic factors while continuing to emphasize economic factors.

At present, foreign countries have already established Neuro-economics and Neuro-finance, finance, etc. which, though at their infancy deserve our attention. Such a kind of situation enables simulation analysis to find new theoretical basis and display its strength.

There are two kinds of interpretations of simulation analysis. In a narrow sense, simulation analysis refers to Experiments with Economic Principles represented by V. L. Smith. In a broad sense, simulation analysis refers to not only experimental economics but also systematic dynamics, ASPEN, SWARM, etc. In other words, all economic simulation operations conducted by computers are within the scope of simulation analysis. The biggest problem facing simulation analysis now is the use of hypothetic data to study the artificial world. Such kind of study is distanced from reality and is of little use. Therefore, we should turn our research objects from the artificial world to the actual world. Micro-data can be used, but they should reflect reality as much as possible.

We can achieve this goal by following three steps: First, make a macro-control model of China according to the theory of ASPEN; second, use real data to conduct specialized studies on population, production, income, investment, finance, consumption, environmental protection, foreign trade, industry and regions, etc.; third, make a model of the findings of specialized studies.

To conduct simulation analysis, we cannot rely on one single simulation technology. Instead, we should use different technologies to analyze different problems and give a full play to their advantages. Therefore, technological integration is a must for simulation analysis. Apart from these, a series of practical economic problems including the transition of developmental modes, industrial upgrades, economic results, growth rate, inflation, construction of new rural areas, environmental protection and ecological balance, technological progress, insufficient effective residential demands, foreign trade and foreign currency are all what quantitative economics need to study.

Part One

Several Issues to Consider in Quantitative Economics*

Wu Jiapei[†]

Quantitative economic research refers to the research of quantitative manifestations, quantitative relations, quantitative changes and laws as well as quantitative analytical methods of the economy (especially economic-mathematical methods). Marxist economics has always attached a great importance to the quantitative research of economics. During the current socialist modernization, approaching the issue of quantitative economic research is of significance to both the development of economic sciences and the improvement of economic work.

1. In our real life, there exists a phenomenon that merits our serious consideration: On the one hand, people tend to pursue quantity while neglecting quality in their practical economic work; on the other hand, they emphasize qualitative analysis and neglect quantitative analysis in economic research.

On the surface, these two tendencies are paradoxical. In the essence, however, they are consistent because quality and quantity are a dialectical unity. Nevertheless, as time, place and condition change, the transition from quality to quantity and the transition from quantity to quality are not exactly the same in terms of their significance or function. As far as practical economic work is concerned, we need to develop scientific technologies and improve management to enhance quality without failing the quantity

*This article came from the 10th Issue of *Economic Research* in 1980.
[†]Wu Jiapei, Researcher of State Information Center, Doctoral Supervisor and Honorary President of Chinese Association of Quantitative Economics.

requirement. As for theoretical economic research, we need to collect ample data, improve analytical tools, employ mathematical methods and computational techniques, etc. to carry out quantitative analysis on the basis of qualitative analysis. When an economy is an extensive one with economic growth heavily hinged to the expansion of means of production, people tend to focus on quantity rather than quality. When it evolves into an intensive one with economic growth highly dependent on the improvement of the quality and efficiency of means of production, people tend to pay more attention to qualitative research rather than quantitative research. Therefore, we should adopt different approaches to economic issues according to the economic developmental stage. At an extensive economic developmental stage where economic factors are easier to analyze, a general quantitative analysis is enough for us to grasp qualitative attributes. However, at an intensive economic developmental stage where economic factors are interrelated and more complicated, quantitative analysis based on pure experience or observance without an in-depth analysis is far from enough. Therefore, we need a transition from quantitative research to qualitative research, which shows our better understanding of the economy or our research progress from an elementary one to an advanced one.[1] However, this progress is constrained by both objective economic development processes and many other factors.

During the past three decades, China's economic research has scored considerable achievements but there are some limitations in considering the practical research findings. For example: Applied economic research is underdeveloped compared with theoretical economic research, quantitative analysis is inadequate compared with qualitative research; specific research on economic methods and methodology are insufficient compared to general research on pure economic theories. For quite a long time, we have neglected economic management sciences, analytical research of

---

[1]To represent the quality of an object as quantity is an important way and symbol of better understanding of an object. It symbolizes the cognitive development of scientific knowledge from an incomplete and inaccurate one to a complete and accurate one. However, this cognitive development is based on not the separation between quality and quantity but their unity. Qualitative research cannot be conducted without certain observation and analysis of quantity while quantitative analysis relies on qualitative research and contributes qualitative research.

economic statistics and the application of mathematical methods in the economy. Quite a few published articles contained numbers or formulas and theoretical conclusions from economic research which was weak and abstract due to the lack of statistical evidence and could not be put into practice. Some academic work that approached quantitative analysis and employed mathematics based on qualitative research was often rejected as something beyond comprehension. This phenomenon can be accounted for from the following several aspects:

First of all, quantitative research that emerged in the early 1960s in China was destroyed by Lin Biao and the Gang of Four who deemed quantitative research and the application of mathematical methods as heresy and condemned them as elements of revisionism and capitalism.

Furthermore, the long-standing disconnection between economic theory and practice made it almost impossible for researchers to obtain economic data (especially statistics) essential to quantitative analysis or overall analysis in particular.

Additionally, our economic research teams, on the whole, had a poor mathematical background. Many researchers did not understand or simply do not care about mathematical calculations and deductions. Therefore, they neglected mathematical training when teaching economics and failed to give due consideration to mathematics even when teaching statistics.

Finally, the root cause of this phenomenon lies in our undeveloped productivity, science and technologies as well as our extensive national economy, which led to an undeveloped economics that did not necessitate an extensive application of mathematics and computational techniques.

After a shift in the China Communist Party (CCP)'s focus of work, while we were making efforts to achieve the goal of "the Four Modernizations" under the policy of "adjustment, reform, rectification and improvement", it became more and more important to develop economics and reinforce economic research on quantitative and methodological issues in our national economic development. For this purpose, we should pay special attention to three aspects as follows:

First, orientate economics towards the goal of "the Four Modernizations". At the same time, when we develop political economics, productivity economics and other theoretical economics, we must also develop applied economics including economic management sciences, technical economics and quantitative economics. In this way, economics is not only used as a concept to equip and guide workers so that they conduct their work in line with objective economic laws, but also is linked to practice and contributes to practical economic work by helping them formulate concrete economic plans and measures; providing various methods and techniques through specifying and refining economic theories. We should carry out meticulous quantitative analyses of major economic problems such as economic systems, economic structures and technological import to elicit accurate quantitative results. As the cause of socialist modernization progresses, it will be of greater urgency to conduct quantitative analyses. Thankfully, the application of mathematical methods and computers makes it possible to conduct meticulous quantitative analyses. Likewise, the study of objective economic laws should not be limited to an accurate statement of economic laws. Instead, it should further reveal the conditions and mechanisms under which economic laws function; measure their quantitative constraints and manifestations to apply them to the organization and management of social production and improve economic performance.

Second, improve economic management and economic plan management in particular. This requires us to strengthen research on quantitative and methodological issues. Economic management is inherently connected with decision-making and forecasting. Qualitative research alone cannot guarantee a sound decision. Admittedly, we have made appropriate judgments and decisions based on experience in the past when economic issues were less complicated. However, given the growing complexity of major problems in present economic development and the general inexperience of policy-makers, we must rely on science and quantitative methods to conduct a comparison and analysis before making a choice or decision so as to avoid major economic losses. The decentralization of decision-making rights and mutual coordination of decisions also necessitate quantitative demonstration and appropriate economic plans. Since carrying out economic forecasting closely related to economic decision-making and long-term economic planning is exploring the trend of economic development and grasping the developmental orientation and extent of the national economy and its elements, we must pay great attention to research on qualitative and timely forecasting and their correspondent methods. In addition, a better equilibrium calculation (especially the research on and analysis of overall equilibrium) in economic plan management and economic plan selection (especially the compilation of optimal economic plans) is also bound to require an extensive application of mathematical methods. However, at present, both methods of economic planning and methods of statistics, accounting and business accounting need improvement.

Third, the modernization of economic management, especially the effective use of modern management means like computers calls for reinforced research on quantitative issues and mathematical methods. It is true that we cannot simply identify the modernization of economic management with the application of computers. However, there is no denying that their advent and application is an important symbol of the modernization of economic management means which has advanced the modernization of economic management systems and methods. Without computers, the modernization of economic management would be impossible. The information revolution represented by computers is a great industrial revolution. It allows machines to undertake part of mental work for humans. But it is more than a reform and development of computing tools. The application of computers must be based on a more in-depth quantitative analysis of the economy and the research on and formulation of mathematic methods and models. In addition, it must also be based on adequate and reliable economic information (data and documents), without which, it would be extremely difficult to organize and manage modernized production even with the help of advanced computational technology. To obtain accurate and comprehensive economic information and transmit, codify, integrate, process, calculate and analyze it, we need to employ a wide range of statistical and mathematical methods and build a powerful and rational economic information system, compatible with computer networks.

2. A socialist economy utilizes the unity of productive forces and productive relations. They both undergo quantitative and qualitative changes, but generally speaking, they only experience quantitative changes when a qualitative change takes place. A socialist economy is a planned economy based on public ownership of means of production. Under a social economic system, the ultimate goals and interests of enterprises are consistent with those of the society as a whole. Therefore, a socialist economy provides a more favorable social condition than a capitalist

economy for quantitative research and the application of mathematical methods. To conduct quantitative economic research, we should get ourselves prepared in these areas:

First, combine quantitative study and qualitative study in theoretical economic work. The objective basis for conducting quantitative economic research lies in the unity of qualitative and quantitative specification of any economic phenomenon and process. Such kind of unity finds expression in not only economic concepts and categories, but also in economic laws. The transformation of an old quality into a new one is precisely based on quantitative changes. The unity of qualitative and quantitative specification of an economic object requires economic research based on a combination of both qualitative and quantitative analysis to avoid imprecise or even erroneous understanding. All sciences must have qualitative analysis as its basis and recognize its paramount significance to quantitative analysis in that without a prior in-depth qualitative analysis, understanding quantitative change and its law is impossible. Due to the social and historical nature of research objects of economic sciences and the complicated interrelations between them, qualitative analysis is more important. Nevertheless, we should never underestimate quantitative analysis, or neglect its role in deepening our understanding of quality and helping us to achieve a comprehensive understanding of economic objects. In our present socialist economic research, we have a tendency to underestimate quantitative analysis though socialist economic organization, management and planning require us to attach greater importance to quantitative analysis instead of qualitative analysis. Qualitative analysis without corresponding quantitative analysis is meaningless and cannot solve practical economic problems.

To develop socialist economic quantitative analysis, we must rely on the formulation and application of economic mathematical models.[2]

---

[2]An economic mathematical model is a formula or a formula system that represent economic quantitative relationships. For instance, the equation set that represents quantitative relationship between aggregate social product and final product with matrix calculus: $(I - A)^{-1}Y = X$ is an economic mathematical model of sector relationship equilibrium. In this equation, X refers to the vector quantity of gross social product and the output of product in different sectors is its elements; Y refers to the vector quantity of final product and the aggregate national income is its elements. A refers to direct consumption coefficient

An economic mathematical model is a major form of scientific abstraction and synthesis. It is the intermediary of economic theory and economic practice. It simplifies economic practice under the guidance of economic theory, but maintains the affinity with it in some major and essential aspects. We can make use of this "simplification" to understand many different perceivable objects according to their common attributes (Engels, 1972, p. 579), obtain information related to economic practice and conduct in-depth research of it. Therefore, an economic mathematical model must reflect certain economic content and possess certain mathematical forms as well. In other words, to use an economic model is to reflect economic processes and relations with mathematical forms. An economic mathematical model can serve both as a tool of theoretical research and a solution to concrete economic problems.

Like other quantitative analyses, the formulation and application of an economic mathematical model must be based on an in-depth qualitative analysis and a full revelation of the nature of economic concepts, categories, laws, economic ties and interdependent relations between economic factors. For instance, in our research of the socialist economic structure, if we do not have a good understanding of concepts such as "sector", "consumption", "appropriation" etc. and categories like "final products", "middle products", "productive technical relations" and "social economic relations" as well as the essential relationships between developing productivity and satisfying demands within basic socialist economic laws, especially the differences and relationships between practical social demands, controllable and affordable demands and market demands subjected to price change and supply–demand condition and the quantity and structure of residential consumption needs, it will be very difficult for us to formulate and apply the equilibrium economic mathematical model based on a balanced relationship between economic sectors. Thus, we should emphasize and conduct quantitative research and analysis of the socialist economy on the one hand and strengthen qualitative research and analysis of economic theory on the other hand. That the socialist economic issues have not been

---

and the direct consumption in one sector in relation to that in another is its elements; I refers to unit matrix. All its elements in main diagonal are 1 while other elements are 0. It functions in a matrix calculus the way 1 functions in an arithmetic.

studied adequately is by no means unrelated to our insufficient exploration of their qualitative attributes.

Second, improve and develop the groundwork regarding economic documents and data. It is impossible to conduct quantitative economic research without ample and reliable economic documents and data. Over the past three decades, national departments, regional governments and SDPC (State Development Planning Commission) and National Bureau of Statistics and other comprehensive authorities have accumulated a large body of precious historical documents about economic development. Moreover, since there is no such thing as "commercial secrets" between enterprises under the socialist system, we can obtain comprehensive documents across the country through the statistic-reporting system. Nevertheless, our groundwork regarding economic documents and data is still insufficient. Worse still, a large number of them were destroyed in the last 10 years. Problems such as messy data, incomplete documents, unreliable figures, non-identical index, three accounts (business account, financial account and statistical account) and divergent data on planned economics needs to be addressed in an earnest manner. To meet the demand of quantitative research, we can improve our groundwork regarding economic documents and data in these ways:

(a) Sort out existing historical documents and data and improve their utility. While obeying certain confidential rules formulated by our country, we can open our economic documents and data to the public and enable them to improve our theoretical research work and serve the cause of "the Four Modernizations".

(b) Improve financial and statistical accounting and make better use of financial documents, statistical documents in particular. The initial recording and calculation of cost and production expenses needs improvement. Statistics must serve not only economic planning but also economic management and research. The current situation where statistics are hooked to economic planning needs a change. To be specific, we should give autonomy to statistics and establish statistical law to give full play to its supervisory role.

(c) Promote the use of partial sample survey's and one-off general survey techniques, as these are effective ways of obtaining major documents. We should stop the old practice of neglecting general surveys.

(d) Promote a unified economic calculation. We should unify as much as possible, i.e. business accounting, financial accounting, statistical accounting, planned calculation and technical economic calculation without failing the special requirements of certain specific calculation, organize economic information in a more scientific way and gradually establish a unified and modernized economic information system.

Third, formulate and promote economic mathematical methods. Methods are ways of understanding and changing reality. Proper methods play a crucial role in scientific research. A mathematical method is an important way of understanding, which is called a dialectical auxiliary tool and means of representation (Engels, 1972, p. 357). It is a specific way of understanding compared with materialistic dialectics as a general way of understanding, but a general way of understanding in relation to various sciences approaching forms of movement. The fact that research objects of economics are similar and comparable is the prerequisite for the application of measurement calculation and mathematical methods. Although many mathematical methods applied in theoretical economic research and practical economic work such as vector-matrix approach for calculating complete consumption, mathematical programming approach for selecting optimal plan, various statistical and mathematical approaches for predicting economic development trends and other methods such as plan evaluation and review techniques used in economic management and regression analysis for analyzing the relationship between economic factors are readymade and effective, once they are popularized, new problems will arise and require to be addressed through mathematical means. A case in point is the proposal to use a goal programming[3] approach after linear planning was applied to economics. Therefore, to conduct quantitative economic research, we need not only economic theory as a guide and ample economic documents as the premises, but also applicable and handy mathematical methods as tools.

---

[3]Goal programming is an extension or generalization of linear programming to handle multiple, normally conflicting objective measures.

In terms of conducting economic quantitative research, I suggest that we should first, forsake our traditional family-style research. Instead, it is the duty of the supreme economic organs or competent authorities to mobilize the strength of research institutions, institutions of higher learning and economic organizations concerned, and attract experts and scholars from various fields to conduct comprehensive multidisciplinary research in a collaborative way. This is especially the case when it comes to economic-mathematical research, with regard to national economic issues like the optimal planning and management of the national economy and the prediction of the outlook for national economic development. It is impossible for any individual to conduct such research within his/her own professional field. For some major research projects and subjects, we need not only the participation of hundreds of units and thousands of people, but also an organization according to the theory and methodology of systemic engineering. Second, we should rely on both the internal cooperation and collective effort within the economic community and that between economists and mathematicians, statistics experts, planning experts, and computational experts. Quantitative economic research involves various fields. Issues like quantitative research and the application of mathematic methods exist in almost all economic disciplines. Even disciplines like economic history and the history of economic thoughts have to integrate or coordinate with quantitative economic research. Once quantitative economic research is started in all economic disciplines within each research field, it will be easy to obtain and improve a general theory and methodology for quantitative economic research. For large-scale quantitative economic research which exceeds the scope of economic sciences and involves mathematics and computational techniques, the participation and assistance of mathematicians, computational technical personnel and systematic analytical experts is a must. Third, we must establish new quantitative economic majors and produce a new generation of economists well-versed in both the economy and mathematics. All economic students at institutions of higher learning must be required to study mathematics. Students majoring in quantitative economics should also study advanced mathematics and computing. It is time for economists to command and use mathematics.

3. After we conduct quantitative economic research in an all-round way, will there be a new discipline of economics — quantitative economics?

Our answer is affirmative. Certainly, due to the insufficiency of current quantitative economic research, the research objects, main contents of quantitative economics and the relationship between quantitative economics and other economic disciplines are still unclear. But based on current studies, we have some perceptions on its nature and notion.

Quantitative economics is a branch of socialist economics which studies quantity, quantitative relations, quantitative changes in the socialist economy and its laws by means of mathematics and computational techniques under the guidance of Marxist economic theory and based on qualitative analysis. It takes all issues related to quantity in the socialist economy as its research objects. Quantity in the economy is a comprehensive notion. It is more than an economic numerical value. According to Marxist economics, quantity involves not only the existence of an object and the speed, degree, scale and form of its development, but also the "rapports"[4] (Engels, 1972, p. 202) between different objects, the structure formed by its components and models that reflect quantitative relations of an object.

The general contents of quantitative economics include theoretical quantitative analysis and experimental numerical value calculation; economic-mathematical research of microeconomics (small-scale economy) and the macro-economy (large-scale economy); application of mathematical methods and computing and other related emerging sciences such as operational research, control theory and systematic engineering in the socialist economy. The goal of establishing and developing quantitative economics is to gain a more comprehensive understanding of the development laws of the socialist economy through economic-mathematical research, so as to improve socialist economic planning and management. Therefore, quantitative economics directly serves socialist economic planning and management.

The major contents of quantitative economics include: Quantitative economic research theory and methodology (especially theory and methodology used for establishing and applying various economic mathematical models), economic-mathematical analysis of socialist expanded reproduction, research on the equilibrium of sector relationships, optimal national economic planning and management, research on economic prediction

---

[4]According to Frederick Engels, rapports refer to quantitative relationships.

and prediction methods, research on economic-control theory, economic calculation in price formation, assessment of investment results and argumentation of investment plans, optimal use of resources, and economic quantitative issues such as production layout, transportation, national reserves, labor force, commodity circulation and civil consumption.

From its contents, we can see that quantitative economics is a multi-disciplinary and integrated marginal discipline. It combines political economics, economic planning, statistics, mathematics and computational science; some contents overlap with quantitative research of other economic disciplines. Therefore, we should tailor quantitative economics to meet the demands of socialist economic development. Meanwhile, we should find out the common characteristics of other economic mathematical research specific to each economic discipline and enrich the general theory and methodology of socialist quantitative economic research to promote its own development and offer guidance to other economic mathematical research.

Quantitative economics is a practical science and is a science of methodology. It takes political economy and productivity economics as its theoretical basis while maintaining a close relationship with sector economics and professional economics. It maintains an even closer relationship with technical economics. In fact, many of their contents intersect and overlap each other. For instance, the rational allocation of processing tasks by machine load, determination of optimal quantity of production and compilation of optimal transport plans and other issues in a small-scale economic study and the optimal use of resources, evaluation of the investment effect and demonstration of investment issues of the national economy in a large-scale economy are not only the contents of quantitative economics but also the subject of technical economics. However, quantitative economics is different from technical economics in that it involves not only quantity problems relating to productivity, but also problems associated with production relations. Besides, when studying quantity problems relating to productivity, technical economics focuses on small-scale economic problems (such as enterprise, sectors and individual productive factor) and the assessment and demonstration of technical policies, plans and economic measures whereas quantitative economics focuses on large-scale economic problems (e.g. integrity of

Three Issues in Quantitative Economics[*]

Zhang Shouyi[†]

In March 1979, the First Colloquium of the Chinese Society of Technology Economy (CSTE) was held. A group of 18 experts who participated in this colloquium and discussed issues related to the application of advanced mathematics and computing to economic research and management. They agreed that "economic-mathematical method" should be the proper name of the research field, but not of the discipline (or a school). As for what the new name should be, opinions differed vastly. Out of the dozen of names initially proposed, three were most favored: Econometrics, economic systematology and quantitative economy. Finally, quantitative economy was adopted as the new name of this discipline which was further changed into quantitative economics in 1984.

It is by no means an easy task to name a discipline or a school which involves a series of issues such as research objects, methods, contents and systems. Opinions can be divergent for a long time. For instance, although statistics is a fairly old discipline, it is still undecided whether it is one discipline or several disciplines and disputes remain on its research objects. Given the youth of our quantitative economics, many problems are to be further discussed.

---

[*]This article came from the 3rd issue of *Jinyang Journal* 1982.
[†]Zhang Shouyi, Honorary Member of Chinese Academy of Social Sciences, Researcher of Institute of Quantitative Economics and Technical Economics, Doctoral Supervisor, honorary President of the Chinese Association of Quantitative Economics.

1. Quantitative Economics is the Mathematical School of Marxist Economics

What is quantitative economics? There are three major views in circulation at present.

According to the first view which is known as "metrical approach", quantitative economics is econometrics. Experts holding this view refuse to name the discipline quantitative economics. Instead, they propose to use econometrics to refer to this discipline whose research objects are quantitative relations between economic objects. We argue that Chinese quantitative economics is a much larger notion than western econometrics in that it includes not only econometrics but also many other disciplines.

According to the second view which is known as "disciplinary approach", quantitative economics is a discipline. Experts holding this view think quantitative economics is to study quantitative performances, relations, changes and laws of the socialist economy by using mathematical methods based on qualitative analysis under the guidance of Marxist economic theory. The merit of this view lies in its clarity in research objects and methods. But we found two problems with this statement: On the one hand, it takes quantitative economics as a single discipline; on the other hand, it only studies the socialist economy. Let us begin with the second problem.

As far as we know, Goldberg, Lawrence Liu and Gregory Chow from America and Shinichi Ichimura from Japan all developed econometrical models of China; Niwa Chun Hei from Japan made an input–output table of China in 1956; in July 1980, Wayans from America announced the development of an input–output table of China between 1952 and 1980. In recent years, the World Bank made input–output tables and an econometric model of China. Likewise, Chinese experts will develop econometrical models of America, Japan and other countries as well as a global econometrical model. Obviously, these models which approach issues outside the socialist economy also belong to quantitative economics.

Now, let's look at the first problem. At present, mathematical economics, econometrics and economic optimization as well as economic prediction and operational decision-making have all developed into independent disciplines with their own characteristics. Therefore, it does not make sense if one of them is considered to include others.

It should also be pointed out that this statement is self-contradictory. On the one hand, these experts presented a series of research topics for quantitative economics such as economic computing of price formation, evaluation of investment effects, demonstration of investment plans, optimal use of resources, energy supply and demand model, production layout, transportation, national reserves, population, labor, commodity circulation and consumption. On the other hand, however, they argued that other economic disciplines study specific quantities within their own sphere while quantitative economics studies general quantity in economics. In fact, all the above-mentioned topics are the unity of quality and quantity and are specific economic quantities. But they are different in nature. For instance, quantity of population is different quantity of investment. As a matter of fact, all these topics are research objects of a certain economic discipline like price theory, capital construction economics, resource economics, energy economics and energy technology economics, production layout theory, transportation economics, supplies economics, population theory, labor economics, commercial economics and consumption economics, etc. From this perspective, quantitative economics, which does not have its own research objects, is not a legitimate discipline. But it has general economic quantity. For example, all the topics mentioned above contain economic quantity that change constantly. For this reason, we believe that if this view is correct, quantitative economics will only study the universality of all economic disciplines related to quantity or a few abstract principles rather than economic quantity itself. In other words, quantitative economics will be hollow and is of no help to practical economic management.

According to the third view which is called "school approach", quantitative economics is the mathematical school of Marxist economics. We are for this view, as it is to conduct quantitative analysis on the basis of qualitative analysis and is one of the traditions of Marxist economics. In his ground-breaking work, *Capital*, Marx made a quantitative analysis of almost all categories and laws, reproduction process, productive price and average profit law in particular. Lenin further made quantitative analysis of social reproduction processes and developed Marx's reproduction theory, in his works such as: "On the So-Called Market Problems", "The Development of Capitalism in Russia" and "A Review on Economic Romanticism". Chairman Mao repeatedly highlighted the importance of "quantity". He pointed out that many comrades neglect quantitative aspects of an object, basic statistics and percentages, and the quantitative limit that determine the quality of an object. Since they do not have the notion of 'quantity' in mind, they cannot avoid making a mistake (Mao Zedong, 1969, p. 1332).

However, after the founding of the People's Republic of China, for various reasons, China's economics were basically limited to qualitative analysis while quantitative analysis were seriously neglected. Meanwhile, since we adopted a planned economy, we made long-range programs and compiled long-term, mid-term and short-term plans. As far as making economic plans is concerned, qualitative and directional analyses, although important, fall far short of being a problem solver. Instead, quantitative analysis and thereby circling out quantitative limits for future economy are badly needed during setting up economic plans. This is also true for other aspects of economics. Due to the negligence of and the call for quantitative analysis in economic plans, it is an imperative and urgent task for us to establish and develop quantitative economics.[1]

Apart from socialist economic issues, quantitative economics also studies economic laws of capitalism. This school inherits and develops Marxist economics. It inherits two things of Marxist economics: (1) It adopts the scientific system of Marxist economics including issues studied by classical Marxism–Leninism writers with mathematical methods; (2) it draws upon the stance, viewpoints and methods of Marxism–Leninism to address problems that occur during the establishment and development of mathematical economics. It develops Marxist economics in three aspects: (1) It studies new economic phenomena and develops Marxist economics through theoretical generalization. This is the common task of all Marxist economists including quantitative economists; (2) it systemizes and standardizes Marxist economics and put forward a series of theories, principles and axioms; (3) it uses advanced mathematics and computing to study and solve economic management problems.

---

[1] Since Chinese economy has transited from a planned economy to a market economy, it is more important than ever to develop quantitative economics.

The application of mathematics in economic research is as old as theoretical economic research. As early as in the 17th century, William Petty from Britain employed "figure, weight and measure" to explain economic problems in his work *Political Arithmetic*. He was reputed by Marx as founding father of modern political economics (Marx & Engels, 1972a, p. 271). In other words, although it professes to study capitalist economics, it actually studies relationship between objects instead of relationship between humans so as to prove the eternality of capitalism. In this regard, it is no different from vulgar economics which does not use mathematics. It contributes to the development of western economics by putting forward or studying local equilibrium theory and all equilibrium theory. Besides, it takes advanced mathematics (mostly calculus) as major or even the only method of research.

The modern capitalist mathematical school grew out of the old mathematical school. The relationship between them is that the theoretical basis of the modern mathematical school includes not only what it inherited from the old mathematical school since the 19th century, such as factors of production theory and the theory of marginal utility, but also new theories such as Keynesianism, Monetarism, rational expectation and the supply-side economics that came into vogue more recently. In terms of mathematical methods, apart from calculus, it includes linear algebra, mathematical statistics and some modern mathematical branches. Thanks to the application of computing in particular, it has taken on a new face. One of its major achievements is experimental economics and the most influential and most frequently studied contents within it are econometrics and game theory.

Although classical Marxist writers, used to employ advanced mathematics to express and analyze economic problems and reveal the exploitative nature of capitalism while establishing and developing political economics, the mathematical school of Marxist economics did not appear until two decades ago. During the past, over one century, Marxist economics has long been beleaguered by capitalist economics and has to focus on qualitative analysis. It was not until over 50 years ago that Marxist economic was put into practice. However, during this period, since socialist countries were busy conducting social reforms, they analyzed and solved economic problems mainly through qualitative methods. Besides, it should also be pointed out that the establishment of the first people's autonomous regime, in an underdeveloped agricultural country made it fairly difficult to establish a mathematical school of Marxism economic science. But the rapid development of social modernization and wide application of modern computer technology created objective opportunities for the growth of this school.

The Marxist mathematical school of economics and the capitalist mathematical school of economics are different in several aspects. First and foremost, the fundamental difference between them does not lie in the use of advanced mathematics and computing in the Marxist mathematical school of economics, but in their strikingly different economic theories. Based on dialectical materialism and historical materialism, Marxist economics has profoundly revealed the objective economic law of social development, proved the historical necessity of the replacement of capitalism by socialism and pinpointed the direction and outlook of communism from socialism. Capitalist economics is still what Marx described as "a discipline that studies superficial relationships, which makes plausible explanations to vulgar economic phenomena; vulgar economics only systemizes the hackneyed and pretentious views of capitalist producers, makes them scholarly and declares them as eternal truths" (Marx & Engels, 1972a, p. 98).

Furthermore, their views on the relationship between quantitative analysis and qualitative analysis are different. The capitalist mathematical school of economics denies the dialectical relationship between qualitative analysis and quantitative analysis and sees advanced mathematics as the major or only method of research. It insists that economics is a branch of mathematics or physics. This view is incorrect because all the problems they study are related to human relations. Take allocation theory based on factors of production as an example. Although it attempts to label salary, profit and land rent similarly to labor, capital and land, these purely "physical wages" cannot conceal the confronting relationship between worker, capitalist and landowner, the three major classes in the capitalist society. Therefore, neither the Marxist mathematical school nor the capitalist mathematical school belongs to mathematics or physics. Instead, they both belong to economics.

The Marxist mathematical school believes in the dialectical relationship between qualitative analysis and quantitative analysis, arguing that the former is the premise of the latter while the latter promotes the former. Scholars from this school speak highly of the role of mathematics in economics. The combination of natural sciences and social sciences and the application of mathematics to all disciplines of social sciences is one of the developmental laws of modern science. The application of mathematics to economics can make the quantitative relationship between categories more accurate. For example, seeking differential utility is seeking derivative or total utility. In addition, just as Wieviorka said, "we cannot avoid three kinds of mistakes if we do not apply mathematics to economics. We may elicit wrong law, make false conclusions and neglect the factors that distort the function of the law. However, the working procedure of mathematics helps us to avoid making these mistakes. It forces us to demonstrate the hypotheses with high precision and makes our conclusion almost absolutely correct. Besides, it produces results comparable to practice and thus shows that we can only get approximate answers to problems" (Bliuming, 1983, p. 29). Finally, against the background of modern socialized production, it is almost impossible to solve any complicated economic management problem without using advanced mathematics and computing.

As discussed above, qualitative analysis is the prerequisite for us to successfully apply mathematics to economic theory and management. Since Marxist economics has undergone a long-term research and practice, qualitative analysis of its economic categories and laws have been advanced enough for the application of mathematics. "The mathematical school of economics is, in its technical sense, a derivation of Marxist economics" (Bliuming, 1983, p. 83).

From our point of view, any discipline that uses advanced mathematics and computing to study economic–quantitative relations lies within the scope of quantitative economics. The keyword in our statement is "study". Those disciplines that merely describe objective economic processes with figures do not belong to mathematical school. Only those who study economic issues and elicit new conclusions with the help of mathematics can be included into mathematical school. For instance, by employing advanced mathematics to study technical economic relations between sectors, input–output theory elicits a new category known as complete consumption which is further divided into complete labor consumption, complete fixed asset investment and complete productive capability, etc. Thus, according to our statement, input–output theory belongs to the mathematical school.

Quantitative economics and modern economic management science have both similarities and differences. Modern economic management science can be divided into two parts: First, to model economic management which means to develop all kinds of economic mathematical models based on qualitative analysis. From this perspective, quantitative economics and modern economic management science are the same; second, to select computer type and series, install stand-alone and terminal and their internet access, which is not an aspect of quantitative economics. Besides, quantitative economics includes mathematical economics that studies economic theories, which, however, is outside modern economic management science.

2. Disciplines within Mathematical School of Economics

Let us begin our discussion on this issue from the relationship between school and discipline. Some scholars think that a discipline can be divided into various schools while a school cannot be divided into various disciplines. We opine that both school and discipline are relative rather than absolute concepts. Lenin remarked that early Marxism is "nothing but one of the numerous socialist factions or thoughts" (1972, p. 437). In fact, apart from socialist factions, there were at that time various non-socialist factions. Lenin also remarked that "the genius of Marx and Engels lies exactly in their development of materialism and one fundamental faction of philosophy during half a century's time" (1972, p. 437). This statement perfectly summarizes the great contribution Marx and Engels have made to economics. Although Marxist economics is only one faction of economics as a whole, there is no denying that it includes various disciplines at present.

Currently, mathematical school includes the following major disciplines.

2.1. *Mathematical economics*

Mathematical economics involves the application of advanced mathematics to theoretical economics (political economics and productive economics), the description of economic categories, processes and laws and the functional relationships between them and the development of abstract mathematical models.

Whether mathematical economics has its own research object or not is still a moot question within western economics. American professor Lancaster from Columbia University (1968) stated that mathematical economics is not a discipline but a research field of economics. British economist Alan disagreed and argued that the research object of mathematical economics is the combination of mathematics and economics (1963, p. 19).

Marxist mathematical economics needs to study economic problems of all social forms, but it focuses on mathematicalizing the socialist part of Marxist economics, proposing a series of theories, principles and axioms and laying a theoretical foundation for the establishment of a mathematical modeling system of the socialist economy. There is a wide range of topics to study within this discipline. At present, we have an urgent need for using advanced mathematics to describe and analyze a few well-acknowledged laws of the socialist economy like the basic economic law, law of planned-proportional development, law of socialist reproduction, law of distribution according to work, law of value, economic returns and time lag, etc. The other task is to employ advanced mathematics to study the categories and laws revealed in *Capital*. This requires us to inherit the invaluable legacy of Marxism–Leninism, deepen and develop it in terms of quantitative analysis based on the new phenomena and characteristics of modern capital and draw on its theories, principles and axioms that are of practical value to our socialist economic study.

Currently, mathematical economics is almost a vacuum in China. We have little literature available and have to read foreign literature extensively. But how to use foreign research results is an issue that should be taken seriously. From our point of view, it is necessary to use some Western economic concepts like elasticity (Marx & Engels, 1972b, p. 474)[2], marginal effects, fixed point, etc., some functions like production function, demand function, consumption function and investment function, etc. and some curves like Engles' Curve, Lorenz Curve, etc. and adopt the Marxist position, view and methods to conduct serious analyses.

---

[2]The concept of elasticity is first proposed by Marx.

2.2. *Econometrics*

Basically, econometrics involves the application of mathematical statistics in the social economy. A cursory reading of any econometrical work suffices to prove this point. All these works include assessment methods of distribution, regression, relevance and various parameters, confidence zones, statistical deduction and assumption test that belong to mathematical statistics. However, when applying mathematical economics that grows out of natural sciences to social economy, we must pay attention to a major characteristic of it. Different from natural sciences, mathematical statistics does not allow for controllable experimentation. This suggests we need to modify mathematical statistics according to the characteristics of social sciences. Therefore, we can say that theoretical econometrics is a form of modified mathematical statistics. It is a discipline of methodology which, compared with mathematical economics, absorbs more of Western econometrics. Applied econometrics is the study of various concrete quantitative relations with all sorts of data, computer and corresponding measurement methods under the guidance of econometrical theory.

2.3. *Economic optimization theory*

Marx's remark "to produce maximum products with minimum labor" (1964, p. 104) sets the requirement for and standard of economic optimization. People in all social systems have a common wish for the most rational distribution and utilization of all sorts of social resources, the greatest access to economic results, the rapid development of social production and improvement of people's life. However, under the capitalist system which has an inherent contradiction between the social nature of production and the ownership means of production by capitalists, this wish can only be realized within enterprises and cannot be realized throughout the society. The socialist system is fundamentally different from the capitalist system. Because of the public ownership of means of production and its ensuing objective economic laws under the socialist system, it is highly possible for us to distribute and utilize social resources in the most rational way throughout the society.

However, possibility is one thing, whereas reality is another. According to our experience during the past three decades since the founding of the People's Republic of China, we need at least two conditions to turn this possibility into reality: To formulate a rational economic management system and to equip our economic management personnel with a set of scientific optimization methods. Currently, our management system is under reform. It will create a prerequisite for the optimal distribution and utilization of resources. As for optimization methods, they are the fastest-growing branch of applied mathematics since the Second World War. Apart from calculus and variation calculation, other methods such as linear programming, integer programming, parameter programming, protrude programming, denting programming, dynamic programming, random programming, queue theory, reserve theory, solution theory, search theory, cost–effect analysis theory, optimal network theory, input–output optimal model and macro-econometrical optimal model were proposed. Since our operational research covers a wider range of topics and issues than Western operation research, economic optimization theory is called economic operation research in China.

It is important for us to continue to study new optimization methods, but more importantly, we need to popularize and apply existing methods extensively, improve management rapidly and achieve real effects. Meanwhile, issues such as the selection of optimization standard and price setting are also to be studied and solved immediately.

2.4. *Economic prediction*

Economic prediction involves making accurate predictions about economic development in the future. For a long time, Chinese economic intellectuals were beset with the view that socialist countries should practice planned economies and stay away from economic prediction which is a peculiar product of capitalism. This bias severely hindered the growth of economic prediction, as a discipline in China and impaired both economic theory and practice.

In the broad sense, economic planning refers to the process of making plans for future economic development. It has a predictive nature, but it is different from economic prediction. Economic prediction has nothing to do with law and does not have to be conducted by a certain department. It usually involves proposing plans based on various factors and possibilities and covers a wider range of issues. Many indexes of economic prediction are more general and less detailed than those of economic planning. Economic prediction is the basis of making economic plans and the implementation and effects of economic plans is an important gauge in testing and modifying economic predictions.

At present, the Economic Planning and Statistics Department of China has established prediction organizations to make predictions. But they are inexperienced and underdeveloped. Therefore, one of their major tasks is to study and draw upon a total of over 30 prediction methods widely in use in foreign countries. Accuracy is the only gauge by which we can test the efficacy of a prediction method. We need to keep improving our existing methods through practice and put forward new and better methods.

2.5. *Business decision-making*

Decision-making is to make decisions. There is a distinction between scientific decision-making and unscientific decision-making. The former refers to making decisions according to decision theory, systems, procedures and methods whereas the latter refers to making reckless decisions without in-depth research or investigation.

Business decision-making includes: Individual and collective decision-making, uni-level and multi-level decision-making, affirmative, risky and non-affirmative decision-making, gain–loss analysis and decision trees, optimal decisions made on the basis of economic models, etc.

The basic contents of the above-mentioned disciplines will change constantly with the development of quantitative economics and new disciplines will appear. As for whether economic systematic theory, economic control theory and economic information theory belong to quantitative economics or not, no agreement has been reached at the moment.

3. The Relationship between Quantitative Economics and Mathematicalization of Economics

We interpret quantitative economics as the mathematical school of Marxist economics not only for the above-mentioned reasons, but also for the relationship between quantitative economics and mathematicalization of economics.

From our point of view, the fundamental task of Marxist economics is to reveal the movement of objective economic laws and use them to reform the world. But it shoulders two specific tasks. On the one hand, it is obliged to disclose the exploitative nature of capitalism, demonstrate the superiority of the socialist society, arm the masses with advanced ideas and accelerate socialist construction. On the other hand, it needs to solve the various management problems of socialist economy which is an extremely large and complicated system where new technologies, products and departments keep emerging and economic ties between enterprises, regions and departments are becoming closer and more intricate.

These two tasks are different. To fulfill the first task, we need to make all efforts to equip the masses with basic Marxist economic theories and raise their ideological and political awareness. The second task, however, mainly targets at economic managers and aims to improve economic management.

Given the dual task of Marxist economics, we propose different requirements for applied mathematics and quantitative analysis accordingly. To fulfill the first task of publicity and mass education, we should focus on qualitative analysis and conduct quantitative analysis on this basis. However, at present, we can only use elementary mathematics rather than advanced mathematics. To fulfill the second task, we should apply advanced mathematics and computing extensively to economics and realize the quantification of economic management based on modernization and automation of economic management. Since mathematicalization of economics refers to a high degree of mathematicalization of economics, it only fits in with the second task.

In this respect, we should draw a lesson from the failure of western economics. Since quite a few western economists chased nothing but mathematicalization of economics and filled their economics books and magazines with symbols and formulas: Western economics is severely detached from the masses and is in crisis. Despite the purpose of vulgar economics to glorify capitalism, the absolute mathematicalization of economics made it incomprehensible to most people (over 90%). Therefore, vulgar economics does not work at all. Considering this, the violent attack on mathematicalization of economics from American economist Galbraith and other western economists is understandable.[3]

---

[3]The School of New Institutionism led by John Kenneth Galbraith went to another extreme by negating downright the application of mathematics in economics.

Marxist economics is scientific and true. But if we cannot differentiate its dual task in an earnest manner, chase overall mathematicalization and pack our newspapers, magazines and books with mathematical symbols and formulas, we will also become detached from the masses and fail to fulfill its task to educate, arm, mobilize and organize the masses. Therefore, we should avoid publicizing an undistinguished mathematicalization of economics which would otherwise lead astray Marxist economics.

In the past, since we failed to draw a clear line between the two tasks of Marxist economics, we failed to explain many problems. For instance, when arguing for the necessity of quantitative economics, we used to quote Marx's remark which suggests that a science cannot be truly perfect until it involves a successful use of mathematics. This well-known remark of Marx has also been cited by foreign economists as an argument for mathematicalization of economics. However, from our point of view, their interpretation of this remark is partial. Given the dual task of Marxist economics, to "a successful use of mathematics" can be interpreted in two ways. In terms of the task of publicity and mass education, economics is truly perfect as long as it uses primary mathematics successfully. However, in terms of the task of management modernization, it cannot be truly perfect unless it has used advanced mathematics and computing successfully.

For another instance, some foreign scholars argued that Marxist economics cannot become a precise science unless it is mathematicalized. This argument was rejected by some Chinese scholars who argued that the Marxist economy had already revealed the developmental laws of human society and Marxist economics confined to qualitative analysis had already become a precise science. We think that both arguments are partial and neither has differentiated the dual task of Marxist economics. From our point of view, different tasks demand different degrees of precision. In terms of publicity and mass education, economics is a precise science as long as it focuses on qualitative analysis and applies elementary mathematics to quantitative analysis. However, in terms of management modernization, economics is not a precise science unless it involves an extensive use of advanced mathematics and computers. Therefore, to be a competent manager in this field, one must be well equipped with advanced mathematics and computing technology.

One more instance. If quantitative economics is interpreted as a single discipline, economics is divided into quality economics and quantity economics. Given the unity of quality and quantity in any object, this division is obviously problematic. This problem will vanish if quantitative economics is understood as the mathematical school of Marxist economics. As afore-mentioned, in terms of the task of publicity and mass education, economics should focus on qualitative analysis and apply primary mathematics to quantitative analysis whereas in terms of the task of management modernization, economics should employ advanced mathematics and computing extensively to conduct quantitative analysis based on qualitative analysis. Therefore, both are the unity of qualitative analysis and quantitative analysis with a different focus.

Bibliography

Alan, R.G.D. (1963). *Mathematical Economics* in Russian translation, 19.

Bliuming, Y.G. (1983). *School of Subjectivity in Political Economics*. Beijing: People's Publishing House.

Lenin, V.I. (1972). *Selected Works of Lenin* (trans.), Vol. 2. Beijing: People's Publishing House.

Mao Zedong (1969). *The Selected Works of Mao Zedong*. Beijing: People's Publishing House.

Mark, K. (1964). *Capital Results of the Direct Production Process* (trans.). Beijing: People's Publishing House.

Marx, K. & Engels, F. (1972a). *The Selected Works of K. Marx and F. Engels* (trans.), Vol. 3. Beijing: People's Publishing House.

Marx, K. & Engels, F. (1972b). *The Completed Works of K. Marx and F. Engels* (trans.), Vol. 23. Beijing: People's Publishing House.

Several Major Theoretical Issues
in Quantitative Economics*

Li Jinhua[†]

The development of quantitative economic theory and methods in China boasts a history of 30 or 40 years, during which, many research findings of theoretical and practical value have been produced. However, up till now, research on major basic theoretical issues such as its research objects and knowledge system and the relationship between quantitative economics and relevant disciplines is insufficient. The reason why these issues are important is that they influence the existence of quantitative economics as an independent discipline and the development path of quantitative economic theory and practice. The reason why they are basic is that all research on quantitative economic theory and methods must begin with a clarification of them and the development and application of quantitative economics is impossible unless these issues are addressed.

Quantitative economics originates from the application of mathematical methods to the analysis of economic phenomena. From the very beginning, people discussed the research objects and knowledge system of quantitative economics, but no agreement was reached. Afterwards, few

*The article came from the 11th issue of *Quantitative Economics and Technical Economics Research* in 2001.

[†]Li Jinhua, Researcher of Institute of Quantitative Economics and Technical Economics at Chinese Academy of Social Sciences, Doctoral Supervisor, Associate Chief-Editor and Director of Editorial Department of Quantitative Economics and Technical Economics Research and Vice-President and Secretary-General of Chinese Association of Quantitative Economics.

people paid attention to these issues. As a result, they remain unsolved until today. Nowadays, with the widespread application of quantitative economic theory and methods to social economic analysis, many institutions of higher learning have begun to offer quantitative economic courses for students, especially postgraduates. Under this circumstance, we have the urgent task of theoretical system construction and discipline construction for quantitative economics. Thus, uncertainty about its research objects and knowledge system will hinder the development of quantitative economics and prevent us from producing advanced quantitative economic professionals. This chapter attempts to summarize and review the arguments and research findings on three issues that trouble the economic community in order to draw more academic attention to these major theoretical questions.

1. Research Objects

In essence, to name the research objects of quantitative economics is to define quantitative economics. Academic views in this regard fall into three major categories: (1) There is no quantitative economics or quantitative economics is econometrics; (2) quantitative economics is an independent discipline; (3) quantitative economics is not an independent discipline but a school of economic sciences.

Some scholars hold that quantitative economics are econometrics and its research objects are the quantitative relationship between economic objects. They argue that the term quantitative economics does not exist in western economics which has econometrics only. In English, there is "econometrics", but there is no specific term for quantitative economics. The term quantitative economics is actually an English translation among Chinese scholars who assume that quantitative economics is Chinese econometrics. Therefore, quantitative economics and econometrics are the same. They both are characterized with quantitative analysis and based on data. According to this view, it does not make sense if we say the Chinese economic community has both quantitative economics and econometrics.

Some scholars argue that quantitative economics is an independent discipline which employs mathematical methods and computational technologies to study the regularity of quantitative performances and quantitative relations and quantitative changes of the economy. But they

energy consumption and economic growth in Malaysia, Singapore and the Philippines; there was a unilateral causality from energy consumption to economic growth in India; there was a unilateral causality from economic growth to energy consumption in Indonesia and there was a bilateral causality between energy consumption and economic growth in Pakistan. Masih & Masih (1997) added the data of China between 1955 and 1990 to their original data. They found that results for the afore-mentioned 6 countries were almost the same and that there was a bilateral causality between energy consumption and GDP in Taiwan, China. Ugur & Ramazan (2003) examined the causality between energy consumption and GDP in 16 countries and found that the series level values of all countries are unstable but their order differences are stable. Furthermore, they noticed a stable linear co-integration between variables of 7 countries. In addition, they found that in Turkey, France, Germany and Japan, energy consumption can boost economic growth. This suggests that the long-term practice of energy conservation in these countries may impair economic growth. The causality between energy consumption and GDP in Italy and South Korea was reversed, while the relationship between energy consumption and GDP manifested itself as bilateral causality.

In 1998, Chinese scholars represented by Zhao Lixia developed a VAR model based on C–D production function to analyze the relationship between energy consumption and economic growth and found a positive correlation between them in China. Lin Boqiang (2001) applied the co-integration error correction model to energy analysis and developed an econometric model of China's energy demand through an analysis of co-integration between energy demand and GDP, energy price and the share of heavy industry in economic structure. Ma Chaoquan *et al.* (2004) adopted the E–G two-stage method to analyze the long-term equilibrium between GDP and total energy consumption as well as that between each component of energy consumption (including coal, petrol, natural gas, hydroelectric, etc.) and found that both co-integration and an obvious bilateral causality existed between total energy consumption, coal consumption and GDP. However, they also found that no co-integration existed between petrol natural gas, hydroelectricity and GDP, but petrol consumption was the cause of GDP and GDP was the cause of hydroelectricity. Zhao Jinwen & Fan Jitao (2007) conducted a Granger causality test of data on GDP and total consumption of primary energy that were made into stationary sequence after first-order differential and found a unilateral causality from energy consumption to economic growth. In addition, by developing a nonlinear LSTR model, they also found that economic growth in turn had a strong impact on energy consumption. Hu Yuxiang & He Changzheng (2007) delved into the causality between economic growth and energy consumption of China between 1989 and 2003 and found that GDP was the Granger cause of total energy consumption and coal consumption. In addition, they also employed GMDH to conduct the same causality test and obtained the same results, namely, GDP was the cause of total energy consumption and coal consumption.

3. Data Collection and Processing

To study the interdependent relationship between China's energy consumption and economic growth, we selected the data on China's energy consumption and real GDP from 1978 to 2007. We marked LGDP as GDP, LTEC as total energy consumption and LCOAL as coal consumption (to eliminate heteroscedasticity of data. We selected logarithm of each data sequence. Therefore, data used in this article were sequences of logarithm). In this way, we marked LPE TROLEUM, LNATURALGAS and LELEC-TRICITY as the total consumption of petroleum, natural gas and electricity, respectively. Data on energy consumption came from *Almanac of China's Economy in 2002*, *Almanac of China's Statistics in 2006*, *Almanac of China's Statistics in 2007* and *Almanac of China's Statistics in 2008*. The unit is 10,000 ton of standard coal. Data on GDP comes from *Almanac of China's Statistics in 2008*. It was obtained through a calculation of GDP index and GDP in 1978 with 1978 as the base period. The elicited GDP data were GDP at constant prices in 1978 and the unit was RMB 100 million. All the regression and test in this article were completed by Eviews 6.0.

4. Empirical Analysis

4.1. *Unit root test*

We first used the ADF (Augmented Dickey–Fuller) test to conduct the unit root test of the log series of GDP, total energy consumption, the

Variables	PP test							
	Statistics	Critical points		DW	AIC	SC	Test setting (I,T,B)	Stationarity Markedness : 1%
		1%	5%					
D(LGDP,2)	−5.43	−2.65	−1.95	1.84	−4.52	−4.47	(0,0,9)	stationary
D(LTEC,2)	−4.66	−2.65	−1.95	1.85	−3.95	−3.91	(0,0,6)	stationary
D(LCOAL,2)	−5.00	−2.65	−1.95	1.91	−3.45	−3.4	(0,0,5)	stationary
LPETROLEUM	−10.58	−4.31	−3.57	2.03	−3.92	−3.78	(1,1,26)	stationary
D(LNATURALGAS)	−6.88	−4.32	−3.58	1.92	−2.68	−2.54	(1,1,2)	stationary
D(LELECTRICITY)	−5.22	−3.69	−2.97	1.99	−2.53	−2.44	(1,0,2)	stationary

Figure 1. Stationary Test of GDP, Total Energy Consumption and the Consumption of Coal, Petrol, Natural Gas and Electricity

Notes: I stands for interval, T stands for trend (1 means include, 0 means exclude), and B stands for interval lag term. The lag term is determined by Newey–West estimated by means of Battlett Kernel.

consumption of coal, petroleum, natural gas and electricity and obtained results. However, these results failed the stationary test of the VAR (Vector Autoregression) model. Therefore, we used the PP method to conduct unit root test of each series and obtained results as shown by Figure 1. From Figure 1 which presented the primary stationary of each series, we can see that LGDP, LTEC and LCOAL are second-order single whole, LNATURALGAS and LELECTRICITY are first-order single whole and the original series of petrol is stationary.

4.2. *Granger causality test*

Every time series data regression analysis had an implicit hypothesis underlying it, namely, the data was stationary. Otherwise, all traditional hypothesis test procedure and results based on t, F and X would be unreliable. The Granger causality test was more sensitive to the stationary of variable series. Take the two variables X and Y as an example. Generally speaking, the Granger causality test can be conducted separately according to different cases of series stationary: (1) When both X and Y are stationary, we can use the VAR model to conduct the test; (2) when X and Y are not stationary but co-integrate, we can use the VEC model to conduct the test; (3) when X and Y are neither stationary nor co-integrate, we can differentiate series and turn them into stationary series and use the VAR model to conduct the test (Chen & Zhang, 2008). What's noteworthy is that, the economic implication of variables now has changed. In the above-mentioned unit root test, we found that horizontal variables LGDP, LTEC and LCOAL were second-order stationary series, namely I (2) series, LNATURALGAS and LELECTRICITY were first-order stationary series and only LPETROLEUM series was stationary. Therefore, we could only test the Granger causality between economic growth and energy consumption (including the consumption of individual energy) under the (2) condition. Since LGDP, LTEC and LCOAL were second-order single whole, there was no point explaining their economic implication after their series were differentiated and test under (3) condition was of no practical value.

4.2.1. *Co-integration test*

This essay adopts the Engle–Granger Two Step method to test the co-integration between total energy consumption, individual energy consumption and GDP. Since LGDP, LTEC and LCOAL were all second-order single-integer series, we could test the co-integration between them.

(1) Co-integration test of GDP and total energy consumption. According to the definition of co-integration, we used OLS to do regression of LDSP and LTEC and get their co-integration regression equation as follows:

$$LGDP_t = -11.92 + 1.85LTEC_t + \varepsilon_{1t}$$

$$(-17.16)\ (30.90)$$

$$R_2 = 0.9715 \quad DW = 0.2163$$

We also employed an ADF test to conduct the unit root test of residual series. Lag order was selected automatically by SIC and the maximum lag term was set one-by-one from 1. Meanwhile, we selected the lag order that could minimize the SC value (thus eliminating the possibility of minimum SC value caused by excessive lag). The results are shown in Figure 2.

LGDP–LTEC						
Regression equation setting	ADF	1% Critical point	5% Critical point	10% Critical point	Test setting (C, T, L)	Conclusions
Terms with interval but without time trend	−2.9	−3.71	−2.98	−2.63	(1, 0, 3)	Interval term not marked
Terms with both interval and time trend	−3.77	−4.36	−3.6	−3.23	(1, 1, 3)	Interval term not marked
Terms with neither interval nor time trend	−3.15	−2.65	−1.95	−1.61	(0, 0, 1)	stationary

Figure 2. Unit Root Test of Residual Terms

Notes: C stands for interval, T stands for trend (1 refers to include and 0 refers to exclude). L stands for lag order. The selection of lag is strictly in line with the principle of SIC. By selecting different maximum lag order, we minimized the SIC value. After the test, we find that the ADF test of residual term co-integration generally does not include trend.

From Figure 2, we can see that a marked co-integration exists between China's economic growth and energy consumption from 1978 to 2007.

(2) Co-integration test of GDP and coal consumption. We adopted the same method to conduct a co-integration test of LGDP and LCOAL and built their co-integration regression equation respectively as follows:

$$LGDP = -12.18 + 1.92\, LCOAL + \varepsilon_{2t}$$

$$(-12.54)\ (22.37)$$

$$R^2 = 0.9470 \quad DW = 0.2033 \tag{2}$$

From Equation (2), we saw a marked co-integration relationship between economic growth and coal consumption from 1978 to 2007.

(3) No co-integration between the consumption of petroleum, natural gas and electricity and GDP.

Since LGDP belonged to I (2) series, LPETROLEUM belonged to I (0) series, LNATURALGAS and LELECTRICITY belonged to I (1) series, which meant they were different series, there was no co-integration relationship between them.

4.2.2. *Error correction model (ECM)*

(1) ECM test model of China's economic growth and total energy consumption was:

$$\Delta LGDP_t = \delta + \lambda e_{t-1} + \Sigma_{t-1} a_i \Delta LGDP_{t-i} + \Sigma \beta_j \Delta LTEC_{t-j} + \mu_{1t}$$

$$\hfill (3)$$

$$\Delta LTEC_t = \gamma + \theta e_{t-1} + \Sigma \Phi_i \Delta LGDP_{t-i} + \Sigma \Psi_j \Delta LTEC_{t-j} + \mu_{2t} \qquad (4)$$

Mind you, $e_{t-1} = LGDP_{t-1} - \sigma - \psi LTEC_{t-1}$ (regression error term of co-integration equation).

To begin with, we used a VAR model[1] to select optimal lag order. The five selective standards namely, LR, FPE, AIC, SC and HQ unanimously selected second-order lag as the final lag order of an ECM test model. The fitting result of the model was:

$$D(LGDP) = 0.0667 - 0.0111(LGDP(-1) + 11.92 - 1.85LTEC(-1)) +$$

$$\qquad (3.78) \qquad (-0.2766)$$

$$\qquad 0.7860D(LGDP(-1)) - 0.0116D(LTEC(-1)) -$$

$$\qquad (3.61) \qquad\qquad\qquad (-0.0739)$$

$$\qquad 0.5289D(LGDP)(-2)) + 0.0753D(LTEC(-2)) \qquad (5)$$

$$\qquad (-2.4) \qquad\qquad\qquad (0.3955)$$

$$\qquad R^2 = 0.4699 \quad DW = 1.91 \quad P(F) = 0.0151$$

In Equation (5), error correction terms (LGDP(−1) + 11.92 − 1.85LTEC(−1)), D(LTEC(−1)), D(LTEC(−2)) were not marked compared with the standard markedness level of 5% whereas constant terms, D(LGDP(−1)), D(LGDP(−2)) were marked. Neither the lag term nor differentiated term of total energy consumption well explained changes in GDP. In addition, the regression parameter 0 before them shows that energy consumption was not the Granger cause of economic growth. Likewise,

---

[1]

$$LGDP_t = C_1 + \sum_{i=1}^{p} \partial_i LGDP_{t-i} + \sum_{j=1}^{q} \beta_j LTEC_{t-j} + \mu_{1t}$$

$$LTEC_t = C_2 + \sum_{i=1}^{m} \theta_i LGDP_{t-i} + \sum_{j=1}^{n} \gamma_j LTEC_{t-j} + \mu_{2t}$$

we could derive another regression result from the same ECM test model:

$$D(LTEC) = 0.0320 + 0.1034(LGDP(-1) + 11.92 - 1.85LTEC(-1)) +$$
$$\qquad (1.91) \qquad (2.12)$$
$$\qquad 0.0692D(LGDP(-1)) + 0.08299D(LTEC(-1)) -$$
$$\qquad (0.4716) \qquad\qquad (4.13)$$
$$\qquad 0.2918D(LGDP)(-2)) + 0.0098D(LTEC(-2)) \qquad\qquad (6)$$
$$\qquad (-1.88) \qquad\qquad (0.0495)$$
$$\qquad R^2 = 0.5939 \qquad DW = 1.86 \qquad P(F) = 0.0012$$

In Equation (6), P value of T corresponding to each error correction term $(LGDP(-1) + 11.92 - 1.85LTEC(-1))$, content terms, $D(LTEC(-1))$ and $D(LGDP(-2))$ was 0.046, 0.0704, 0.0005 and 0.074. Therefore, under the condition of standard markedness of 5%, our original postulate that coefficient before lag term and differentiated term of GDP was false. So, we accepted the alternative postulate, namely, not all coefficients were 0. For instance, since LGDP (-1) was marked, LGDP could explain or predict LTEC. This showed that economic growth was the Granger cause of energy consumption.

(2) As there was a co-integration between LGDP and LCOAL, we could use the ECM model to test the Granger causality between economic growth and coal consumption. Again, we selected optimal lag order with a VAR model and found that the five standards LR, FPE, AIC, SC and HQ unanimously selected second-order lag. Therefore, we took second-order lag as the final lag order of the ECM model. The fitting result of the ECM model was as follows:

$$D(LGDP) = 0.0683 - 0.0087(LGDP(-1) + 12.18 - 1.92LCOAL(-1)) +$$
$$\qquad (3.75) \qquad (-0.3357)$$
$$\qquad 0.7694D(LGDP(-1)) + 0.0038D(LCOAL(-1)) -$$
$$\qquad (3.66) \qquad\qquad (0.0321) \qquad\qquad (7)$$
$$\qquad 0.5181D(LGDP)(-2)) + 0.052D(LCOAL(-2))$$
$$\qquad (-2.49) \qquad\qquad (0.3781)$$
$$\qquad R^2 = 0.47 \qquad DW = 1.92 \qquad P(F) = 0.0149$$

Regression results showed that error correction terms (LGDP(−1) + 12.18 − 1.92LCOAL(−1)), D(LCOAL(−1)), D(LCOAL(−2)) were not marked compared with the standard markedness of 5% while P value of T corresponding to constants, D(LGDP(−1)) and D(LGDP(−2)) was 0.0012, 0.0015 and 0.0214. Thus, we found that lag order of LCOAL could not explain or predict changes in LGDP which suggested that coal consumption was not the Granger cause of economic growth. Meanwhile, we also obtained another regression equation result from an ECM model:

$$D(LCOAL) = 0.097(LGDP(-1) + 12.18 - 1.92LCOAL(-1)) +$$
$$(2.3)$$
$$0.1419D(LGDP(-1)) + 0.7647D(LCOAL(-1)) -$$
$$(1.00) \qquad (4.72)$$
$$0.0691D(LGDP)(-2)) + 0.0166D(LCOAL(-2)) \qquad (8)$$
$$(-0.5475) \qquad (0.1183)$$
$$R^2 = 0.5382 \qquad DW = 1.91$$

Since constants were extremely unmarked, we omitted them and got the above results. According to the fitting result, p value of T corresponding to error correction terms (LGDP(−1) + 12.18 − 1.92LCOAL(−1)) and LCOAL(−1) was 0.031 and 0.0001, suggesting that they were highly marked while D(LGDP(−1)), D(LGDP(−2)) and D(LCOAL(−2)) were unmarked. In addition, since residual terms in the regression equation were white Gaussian noise series, we found that the lag order of LGDP could explain or predict LCOAL, which means that economic growth was the Granger cause of coal consumption.

4.2.3. *Dynamic trends and forecasting*

We employed Henry's modeling method from the general to the individual to eliminate lag orders with unmarked regression coefficients, adjusted residual terms to meet the demands of white Gaussian noise and got a modified ECM model as follows:

$$D(LTEC) = 0.0347 + 0.1026(LGDP(-1) + 11.92 - 1.85LTEC(-1)) +$$
$$(2.62) \quad (2.27)$$

$$0.8499D(LTEC(-1)) - 0.2570D(LGDP(-2)) + \varepsilon_1$$
$$(7.37) \qquad (-1.77) \tag{9}$$
$$D(LGDP) = 0.1268 + 0.6731D(LGDP(-1)) - 0.6064D(LGDP(-2)) -$$
$$0.4069D(LGDP(-5)) + \varepsilon_2$$

We conducted a series correlation LM tests of residual terms in Equation (9) with second-order lag selected by SIC as optimal lag order of residual terms. The result showed that under the condition of standard markedness of 5%, there was no series correlation in residual series. In addition, the model also went through white heteroscedastic test and showed that heteroscedasticity did not exist in residual series, that no correlation existed between residual and explanatory variables and that model specification was correct. Besides, a Histogram–Normality test showed that residual series conformed to the positive distribution with mean 0. Finally, the forecasting test of Chow proved that the model was highly stationary.

We developed a model from Equation (9) to make a dynamic forecast of China's annual GDP and annual total energy consumption in the next few years up to 2020. The results are shown in Figure 3.

To modify our predictions based on the actual observations of 1978–2007, we can control the error of dynamic prediction within 5%. Given that a prediction with a relative error rate below 5% is generally considered as a precise one, our prediction is very precise. From Figure 3, we can see that the real GDP (at the constant prices of 1978) and total energy consumption will reach RMB18.37332 trillion and RMB44.73426 trillion tons, respectively. In other words, GDP will rise by 236% from 2007 and

Year	GDP (trillion yuan)	Total energy consumption (10,000 tons)	Year	GDP (trillion yuan)	Total energy consumption (tons)
2009	6.45981	25.39138	2015	11.46627	34.55353
2010	7.11027	26.74613	2016	12.58027	36.37056
2011	7.83570	28.17372	2017	13.80982	38.30076
2012	8.63436	29.66357	2018	15.17561	40.34388
2013	9.50445	31.21565	2019	16.69317	42.49131
2014	10.44621	32.84041	2020	18.37332	44.7342

Figure 3. Forecast Results of GDP and Total Energy Consumption

total energy consumption will rise by 68% from 2007, an annual average growth of around 5% (this result is consistent with RICS energy report result). We can also derive a general trend of energy consumption from this dynamic trend. The general trend shows that China's energy consumption intensity in the future will keep falling. However, according to the dynamics of energy consumption during the Eleventh Five-Year Plan period, if the dynamics of economic growth and energy consumption remained the same as before, by 2010, the decrease rate of energy consumption in relation to that in 2005 will fail the restrictive index of 20% proposed in the Eleventh Five-Year Plan. Therefore, in the next a few years of the Eleventh Five-Year Plan period, we should make greater efforts to reduce energy consumption and make more efficient and scientific development policies.

Likewise, we used Hendry's modeling method from the general to the individual to eliminate the lag terms in Equation (8) with unmarked regression coefficients, adjust residual terms to meet the requirement of white Gaussian noise and used the modified model to make dynamic forecast of the annual GDP and coal consumption in the next a few years until 2020. The results showed that by 2020, China's coal consumption would increase by 90.75% compared with 2007 and the average annual increase would reach 6.98%, far above the growth rate of total energy consumption.

5. Conclusions and Suggestions

This essay adopts the co-integration test and ECM to conduct an empirical study on the independency between China's economic growth and energy consumption between 1978 and 2007. It makes the following conclusions and suggestions:

(1) During this period, there exists a unilateral causality from economic growth to total energy consumption and coal consumption. To be more specific, economic growth is the Granger cause of total energy consumption both in the long run and in the short run and it is the Granger cause of coal consumption only in the long run. However, neither total energy consumption nor coal consumption is the Granger cause of economic growth. This suggests that energy as a factor of production is not the determinant of economic growth and the reduction of energy consumption

strategy s_i is called a dominant strategy.[1] If every participant's strategy happens to be an optimal response to strategies of other participants, namely, $u_i(s_i, s_{-i}) \geq u_i(s_i, s_{-i})(Vs_i' \in S_i)$, then strategy portfolio $s = (s_i, s_{-i})$ is a Nash equilibrium and $u_i(s_i, s_{-i})$ is the Nash equilibrium utility of i.

The establishment and development of game theory and the formation and refinement of the concept of Nash equilibrium are closely and multi-dimensionally related to human behaviors. However, as a theory that mainly approaches strategic behaviors, game theory derived its thoughts mainly from the three schools of thoughts and theories. The first origin is philosophical thoughts, such as the traditional Chinese art of war, the traditional Chinese culture of harmony and Western structuralism which stresses the mutual influence of strategies. However, for lack of economic background and scientific analytical tools, some enormously valuable ancient thoughts were lost. The second origin is research conducted by Cournot and other scholars on oligopoly and the incomplete competition problem in the modern market economy in particular, which, thanks to the development of scientific analytical tools and methods like mathematics and computers, helped set up and enrich game theory based on the neoclassical economic hypothesis of rational man. The third origin is Marxist's historical materialist and dialectical materialistic views on human duality (naturalness and sociality) and method of class analysis (conflicts between interest groups), labor value theory in classical economics (behaviors determines results) and biological evolutionism. These views all tremendously influenced the theoretical basis, elementary methods and the development direction of game theory. The new advances in game theory and new characteristics of modern economic development highlighted the influence of these views.

1.2. *Refinement of Nash equilibrium*

Game theory which takes interactive strategic behaviors as its research object is the quintessence of human wisdom. It undergoes constant

---

[1]For convenience's sake, omit weak dominant situation where $s_{-i}(V_{s-i})$, if $u(s_i, s_{-i}) \geq u_i(s_i', s_{-i})$.

refinement and improvement against the background of the modern market economy through twists and turns. The core of game theory is the Nash equilibrium. It develops along a clear path: first, the concept of the equilibrium solution of a strategy game was put forward by Nash (1950); then, backward induction was adopted to obtain the sub-game refinement equilibrium solution of extensive and dynamic game models (Selten, 1965); later, type and probability distribution were used to establish game models with imperfect information and the concept of Nash–Bayesian equilibrium solution (Harsanyi, 1967–1968). Since imperfect information and multi-stage dynamic game structure (repetition) in a game are complicated, "the equilibrium selection problem is of particular importance" (Harsanyi & Selten, 1988). It affected the application of game theory in actual economic life at a time to such an extent that some people thought it was entrapped by an analytical mud of perfect competition and perfect monopoly (Camerer, 2003). However, there is no doubt that the entire non-cooperative game theory was developed based on the refinement of Nash equilibrium (Fudenberg & Tirole, 1991).

1.2.1. *Sub-game perfect equilibrium*

The refinement of the Nash equilibrium mainly aims to make up for the flaws in the equilibrium concept of Nash. The concept was improved remarkably after game theorists began to consider extensive game trees. An extensive or dynamic game can be decomposed into several sub-games (follow-up games after every single point) and used for modeling games with concrete sequences of action. Extensive game equilibrium specifies every move of every player at every information set. Sub-game perfect equilibrium emphasizes further restriction. In other words, participants do not select their strategies unless they are in a sub-game situation.

1.2.2. *Bayesian–Nash equilibrium*

In extensive games with imperfect information, there is at least one player who does not know the strategy set or payoff function of other players. At the start of the game, we will use the term "nature" to specify the "type" of a player. In this way, the player knows his/her own type but does not know other players' type. Then, we assume that the probability distribution of players' type is common-prior.[2] Compared with the Nash equilibrium, the Bayesian–Nash equilibrium has two more new characteristics: (1) In the equilibrium path (all behaviors that appear with normal probability at an equilibrium point), all players must employ the Bayesian principle to update their beliefs about other players' actions; (2) In the non-equilibrium path (behaviors after what do not appear at an equilibrium point), players should hold certain beliefs about other players' type. This is how the Bayesian principle imposes a minimum restriction on the beliefs about possibility in a non-equilibrium path. In so doing, it excludes the possibility that players may destroy the optimal equilibrium after they observe the behaviors in a non-equilibrium path.

1.2.3. *Trembling hand perfect equilibrium*

Selten (1975) devised a wise method to conform beliefs that deviate from the equilibrium path to Bayes' Rule. This method is known as trembling hand perfect equilibrium. The concept of trembling hand equilibrium assumes that even in equilibrium, game players are likely to make certain mistakes when choosing strategies. In other words, all trajectories in a game tree will have the normal probability of being chosen and Bayer's Rule, therefore, can be used to change some ideas. Trembling hand perfect equilibrium refers to the limit that trembling Bayesian–Nash equilibrium can reach when the tremble probabilities goes to zero. Later, a number of experts further refined trembling hand perfect equilibrium: Kreps & Wilson (1982) proposed sequential equilibrium which is the co-twin of trembling hand perfect equilibrium; Myerson (1978) argued that when there is a huge difference between equilibrium strategy function and non-equilibrium strategy function, trembling probability is supposed to be very small, which, therefore, results in the concept of proper equilibrium. There are also some other refined concepts of signaling game which can take into account both imperfect information and dynamic games.

---

[2]This is called common-prior hypothesis.

1.2.4. *Quantal response equilibrium (QRE)*

In a quantal response equilibrium, players will not choose one probability to make "the best response" (they will do so in a Nash equilibrium). Instead, they will make a "proper response" and choose with high probability the response with a relatively high pay-off function. As a matter of fact, quantal response equilibrium is often defined in terms of logit or index pay-off function. To put it simply, in quantal response equilibrium, players fix their strategy, then form their beliefs about other players' behavior; compute expected payoffs based on these beliefs. To compute the payoff from every strategy is to outline the expected utility of every possible strategy. Then, player i makes choices according to the expected payoff from that strategy and makes a proper response accordingly.

1.3. *Obstacles to the development of game equilibrium and cause analysis*

Human behavior does not always tend towards gambling. However, game theory aids the study of strategic economic behavior. The core concept of the Nash equilibrium is that no player has anything to gain by changing only his/her own strategy. In addition, every player makes assumptions on other players' strategies. It can be said that before the 1990s, standard or traditional game theory which holds that hyper-rational players make assumptions after deliberate reasoning and exclusive specification predominated game theory. However, it failed to address two basic problems facing the application of non-cooperative games: One problem is that we have not yet theoretically understood how players can ultimately reach the Nash equilibrium and the other problem has something to do with multi-equilibrium, namely, why a certain specific equilibrium will ultimately be chosen. Since players have different degrees of rationality, strategic behaviors are complicated. As a result, response function that negotiates the information set of every player is not in a 1–1 correspondence relation. This is the fundamental reason why the game equilibrium solution is not exclusive. The contradiction between rational behaviors of homogenous players and strategic behaviors of players with different degrees of rationality constitutes an obstacle to the development of traditional game theory based on hyper-rational or super-rational hypothesis.[3] It is impossible for the utility function based on individual rationality and the concept of entire optimization to be perfectly consistent all the time. This is the root cause of the non-uniqueness of the Nash equilibrium solution. In other words, $u_i(s_i^*, s_{-i}^*)$ is not always equal to Max $u_i(s_i, s_{-i})$. The difference between them is where the complexity with game equilibrium selection appears. However, if we accept that $u_i(s_i, s_{-i}) = u_i(s_i)$ as long as $i \in I$, we accept the basic behavioral assumptions of neoclassical economics which simplifies subject interactive decision-making into independent individual decision-making. This is precisely the theoretical basis of neoclassical economics.

2. Cooperative Games and Coalition Profit

Cooperative games or coalition games and non-cooperative games are complementary. They are integral part of game theory and constitute two classical frameworks of game theory. A non-cooperative game does not refer to one in which players always refuse to cooperate with other players. Instead, it refers to a game in which there is no rule that requires players to negotiate and communicate with each other and there is no forcible agreement that can guarantee every player's selection of cooperative strategic behavior. Comparatively, a cooperative game is more comprehensive and it advocates, respects and supports the idea of honesty and faithfulness, equal development, cooperation and win–win as well as its corresponding behavioral norms. It can also be seen as an ideal state of dynamic non-cooperative game.

The requirement for the existence of cooperative games is that each individual player gains more profit in a cooperative game than in a simple game (excluding the cost of cooperation). If there is a force, whether it is a binding agreement, rule and moral convention or a special opportunity and condition, which can meet this requirement, it is meaningful to study cooperative games. Non-cooperative games mainly approach individual behavior while cooperative games focus on organizational behavior

---

[3]Roughly speaking, hyper-rational = individual rationality + common prior.

(distribution standard, means and result). In a non-cooperative game, major characteristics of individual behaviors are reflected in many aspects such as under the mutual influence of strategies, what kind of decision does every "rational" player make, how does every "rational" player take action, what will be the game result, etc. However, the major problems with cooperative games, include attaching importance to collective action with equal participation of individuals, how is a coalition (organization) formed and what are the basic characteristics of different coalitions, how to distribute cooperative profits within the coalition, etc. Through in-organizational means of distribution, core value and Shapley value or α solution and β solution are all cooperative game solutions obtained under different standards (like TU and NTU, etc.). There are three ways of approaching cooperative games: (1) Domination distribution strategy, which is to define stability set, core and other concepts of cooperative game solution through domination distribution strategy; (2) Valuation strategy, which is to study the interrelationship, distribution principle and cooperative interests, such as Shapley value; (3) To employ bargaining (or prior-game communication) as a means or tool to achieve cooperation between players. This is to study collaboration within the non-cooperative framework and see collaboration as the result of the optimal individual strategy selection under mutual influence. The former two strategies are within the cooperative framework where negotiation, communication and binding agreement are fixed by prior system. In other words, cooperation is exogenic and its practical significance is that modern companies (or organizations) determine stock equity distribution plans through interior negotiation and determine income distribution principle according to system, sponsor nations or permanent member states of the International Cooperative Organization have preferential rights, universal rights of and rules for general member states and new member states, etc. However, the third strategy is a progressive method which attaches importance to the game process. In terms of the essence of the concept of equilibrium, non-cooperative game equilibrium and cooperative equilibrium are the same. Non-cooperative game mainly approaches the game without the coalition of players. If we are to consider the coalition of players and the distribution of cooperative profits among coalition members, we need to study a cooperative game.

A cooperative game is to study the game from a different perspective and from the inside out. Compared with non-cooperative game, it attaches greater importance to collective effects and describes and studies game equilibrium from a different angle, with a different analytical method and research focus. However, it has some similarities with non-cooperative game in terms of research object and interested issues. On the one hand, through bargaining and other methods, the cooperative game shows a tendency to fit itself into a non-cooperative game; on the other hand, a non-cooperative game can reach a better equilibrium through cooperation and helps achieve equilibrium selection. The cooperative game is of particularly enormous theoretical and practical significance to the study on interior organizational management organization, distribution principle, annexation of enterprises, international and regional cooperative organizations, market competition structure, general equilibrium, bargaining, etc. Therefore, the contents of a cooperative game are an integral part of the entire system of game theory. Compared with a non-cooperative game, the development of a cooperative game was slower due to more obstacles to it. Obstacles to its development mainly came from three interrelated aspects:

(a) The formation process of coalition, the stability of the coalition with the entrance and exit of players, the relationship between each individual player's payoff function and residual distribution quota of cooperative value.
(b) Challenges posed by different theoretical views to the domination view within coalition game framework.
(c) Demonstration of the concept of game equilibrium solution and the increasingly vaguer dual explanation of norms (Montet & Serra, 2003).

However, the fundamental problems revealed by both cooperative and non-cooperative games as well as the integration of cooperative and non-cooperative games are being tested, solved and proved amidst the gradually developed evolutionary game theory and game experiments.

It would be ridiculous to describe biological behaviors in nature such as the reproduction and evolution of biotic species with absolute rationality, but they did make "game" equilibrium selections, and conformed to the rule of Natural Selection. Besides, there is still no unified analytical model, equilibrium concept and solution method to deal with cooperative games.

In this light, most game theorists and economists nowadays have shifted their interest to evolutionary game models and game experiments. Therefore, we should have noticed that traditional game theory is based on hyper-rationality, but it does not mean game theory must be totally based on rationality.

3. New Advances in Game Theory

Since the 1980s, new advances and major breakthroughs have been made in game theory.

3.1. *Evolutionary game theory and game learning theory*

Not all game players in real life are rational and even if they are rational, they are unlikely to have the same degree of rationality. Therefore, game theory is evolutionary and needs to be learned in different ways. The primary task of evolutionary game theory research is to determine how an individual in an interactive decision-making situation adjusts his or her behavior according to others' behaviors. This is a key element of an overall structure. The fact is, however, we do not have sufficient data on how an individual makes a decision. Therefore, we have to rely on daily observations and some experimental data to make some seemingly plausible hypotheses. Evolutionary game theory and game learning theory are closely related to each other. Some widely-discussed basic types of evolutionary game learning behaviors and their adaptive mechanisms are detailed in the following sections (Young, 1998; Weibull, 1995):

3.1.1. *Repetition or natural selection*

Compared with those who choose strategies with high payoffs, people who choose strategies with low payoffs are more likely to repeat their strategies. Therefore, people choosing strategies with low payoffs will decrease in a long run. This kind of standard game learning model is a basic evolutionary learning model and is called replicator dynamics or RD Model.

3.1.2. *Imitation*

Imitation or infection describes the behavior of imitating or being imitated as well as the influence of others on an individual's learning attitude, especially those popular or apparently high-yielded behaviors. It may be purely driven by the popularity of the behavior or "inertia" or the arguable correlation between yield and the tendency of imitating or being imitated. For instance, people may imitate the first person they meet. The possibility of their imitation has a negative correlation with their own yield but has a positive correlation with the person they want to imitate.

3.1.3. *Reinforcement learning and elimination*

People tend to adopt behaviors that produced high payoffs before and avoid behaviors with low payoffs so as to reinforce behaviors that are beneficial to themselves. Otherwise, they would be eliminated. Reinforcement model is a standard learning model in behavioral psychology and is attracting more and more attention from economists. In the imitation model, payoff describes selection behavior, but it only pays attention to one's own previous payoffs, rather than others' payoffs. Therefore, its premise is that the possibility of adopting a present behavior at present increases as the payoff from a previous behavior.

3.1.4. *Best strategy and belief learning*

Belief learning model describes the behavior of specifying people's expectation of others' behaviors and optimizing their own expected payoffs. This method includes many learning rules and assumes that people have different degrees of "rationality" or "complexity" when predicting others' behaviors. In this learning model which is the simplest of all models, people choose the best strategy according to the experimental distribution of their opponents' previous behaviors, known as "virtual game". Certainly, there are many other complicated rules on learning behaviors, according to which people adjust their beliefs about others' behaviors.

After Biologist Maynard Smith (1974) first advanced his core idea of ESS and basic expressions in an evolutionary game, Van Damme (1991) constructed a non-intergeneration overlapping model of dynamic genes replication and proposed SSE when studying widespread symmetrical and coordinated games. Binmore (1992) and Samuelson (1997) built an intergeneration overlapping model which stimulated the enthusiasm of people to apply evolutionary ideas and RD model to the economy and trade. A visualized explanation of an evolutionary model is that in capital market or investment activities, when the expected payoff from a certain investment activity or a certain stock is higher than average, the investment activity or stock is likely to attract more people. This creates an accumulation effect which influences, determines and changes the pattern of a capital market in a certain situation. One of the important changes is that we do not have to assume that investors or stockholders are highly rational. Instead, we pay attention to their experimental practices or actual behaviors of following the crowd or conforming to certain conventions.

3.2. *Game experiments*

Game theory since its establishment has never freed itself from difficulties in collecting empirical data for the reason that it aims at research of strategic behaviors, so it is characterized by being rich in norms but short of real evidence. Norm analysis is rigorous in theoretical logics, although the ultimate standard to testify truth is practice. Thus it is necessary to figure out how to prove the validity of research results through experiments and carry out a controllable laboratory study. Despite the fact that laboratory research is not a real situation, a specialized experimental design can stimulate real situations properly as well as verify a specific theoretical matter, which deepens our understanding towards game theory and shows us a new way to apply it. In this regard, experimental research of game theory has made progress along with the development of its theoretical study. In particular, the 2002 Nobel Prize in Economics going to behavioral and experimental economists boosted the experimental research of game theory remarkably.

3.2.1. *Principles and basic norms of experimental economics*

There are three clues offered for us to trace the theoretical origin of experimental economics. The first clue is game experiments, which is also the main constituent of experimental economics. The examination of decision making under uncertainty via experiments opened the door of

CAQE, Professor Zhang Shouyi, proposed to set up the Professional Committee of Economic Game Theory and this proposal was later approved by the Executive Committee. In 1998, the National Seminar of Game Theory was held in Beijing and the CAQE Professional Committee of Economic Game Theory was founded at the same time. It was renamed the China Association of Game Theory and Experimental Economics. This association has organized several national conferences and academic exchanges concerning Game Theory. In 2008, it invited American Professor Daniel Houser to China to hold seminars on Game Theory and Experiment. Researchers in universities and research academies have been following closely the newest developments in game theory, and actively apply it to the economy and management. Some achievements are listed as follows: Game theory and finance (Chen Xuebin, 1997); analysis of company management from the perspective of game theory and the decision-making equilibrium model of entrepreneurs (Wang Guocheng, 2002); game theory and economic system simulation (Wang Wenju, 2003); exploration of game experiment and experimental economics (Ge Xinquan; Wang Guocheng, 2006); game theory courses are offered in several universities, and master's students and Ph.D. students are enrolled in this field. All these have contributed to the development of quantitative economics, and to the theory construction of economic theories and their real-world application in China.

4.1. *Cooperative game and the construction of a harmonious society*

The ultimate foundation of constructing a harmonious society is the coordination of economic interests. At present, friction and conflict between various interests and social classes in China are becoming increasingly salient. This fact deserves more serious consideration, as every single interest group is a participant in the social economy. This problem can be solved by applying cooperative game theory to convert pure competitive efficiency notion to the competitive and cooperative efficiency notion. Dominant distribution strategy, valuation assignment and negotiations involved in the cooperation game mentioned before can be applied to analyze these real-world problems and facilitate cooperation among different interests groups. Their application can further simplify the contradictory interest relations to unsymmetrical coordination games. Furthermore, given that the overall national interests, the influence of traditional culture and powerful political guarantee are all present, this can be the coordination equilibrium with a focal point. Therefore, it is necessarily for us to prompt the realization of this coordination equilibrium and to further generate proposals on income distribution reform and related realistic measures, and to ultimately boost the construction of a socialist harmonious society.

4.2. Evolutionary game theory and micro subjects behavioral heterogeneity

There are common characteristics and laws among the elements and basic relations in an economic system, which become the theoretical foundation of homogeneity behaviors. However, because of differences in the influences of social, historical and cultural factors, behaviors of micro subjects might differ greatly. Correspondently, the characteristics of the economy function and micro foundations' reaction towards economic policies may be different. This is similar to the view that the characteristics of different samples have to undergo hypothesis test before we can draw a conclusion as to whether they are acceptable or not. In the peculiar socio-economic situations of our country, it is beneficial and necessary to study the heterogeneity of micro foundation, e.g. their copy behaviors, imitation (contagion), reinforcement, self-adaptation ability, and learning functions. Based on this, a simulation model of complex economic systems can be built and dynamic simulation experiments can be conducted; insights regarding what types of socio-economic structure to construct and how to construct it, and of the characteristics and types of the micro subjects of macro regulation can be generated. In so doing, public management and policies can be more specific, more differentiated, and effective, and the macro regulation ability and management level will be elevated and enhanced.

4.3. Game experiments and the view on fairness and efficiency

It is a significant development trend that contemporary economics emphasizes individual's interactive behaviors and psychological factors. In the real world, psychological activity is bound to affect people's choices and decision-making, and will naturally affect the utility of game theory. Besides, social bibliographies vary from person to person and change with time, situation and context. Psychologically speaking, many fellow citizens, influenced by traditional culture and their specific circumstance, dread inequality and scarcity, value justice and fairness, and are mindful of time, place and harmonious relationships in cooperation, are tremendously grateful for help received, and at the same time, revengeful for ill and unfair treatment. All these psychological traits may affect their economic behaviors, which deviates from the behavioral self-interest determinant assumption. Therefore, adopting game experiment methods to study people's social bibliography like their views of fairness and efficiency, can reveal more about behavior motives and patterns, provide a new perspective to analyze and solve real world problems and more sensible explanations and foresights for economic reform and development, and lay a more solid foundation for the construction of economic theories.

Application and Development of Input–Output Analysis in China

Liu Qiyun* and Xia Ming[†]

1. Introduction and Early Development of Input–Output Analysis in China

American Economist Wassily Leontief's article "Quantitative Input–Output Relationships in American Economic System" published in *Economics and Statistics Review* in 1936 marked the birth of input–output analysis. The establishment of input–output analysis has had significant impacts on subsequent economic theories, analytical methods and policy studies.

Because of the active efforts of Qian Xuesen and Hua Luogeng, the Operational Office of Mathematical Research of Chinese Academy of Sciences set up an economic research team in 1959 and began to investigate input–output technology. Meanwhile, the Economic Institute of Chinese Academy of Sciences also established a research team to study input–output techniques. Members of these two research teams include Li Bingquan, Chen Xikang, Wu Jiaperi, and Zhang Shouyi, etc. who are considered as the first scholars to have introduced input–output analysis to China.

Among Chinese institutes of higher learning, the Faculty of Planned Economy at China Renmin University was the first to investigate input–output techniques under the guidance of Zhong Qifu. Their focus was to

---

*Liu Qiyun, Professor of School of Statistics at China Renmin University and Doctoral Supervisor.
[†]Xia Ming, Professor of School of Statistics at China Renmin University.

begin with input–output analysis and proceed to explore the application of modern scientific methods and quantitative economic and analytical techniques in planned statistical work (Zhong, 2007).

After the Cultural Revolution was launched, research work was interrupted and the research and application of input–output techniques was arrested. In fact, almost all projects on the research and application of input–output methods in China were stopped at that time. In such a difficult situation, State Development Planning Commission took the advice of Chen Xikang and other scholars to compile an input–output table. Under the approval of the State Development Planning Commission, scholars represented by Chen Xikang cooperated with major units including Beijing Institute of Economy, Chian Renmin University and the Computing Centre of State Development Planning Commission for 2 years from 1974 to 1976 and managed to compile an input–output table of 61 major products of China in 1973. In 1979, this table, which was considered to be the first of its kind, was turned into a book by the Computing Centre of State Development Planning Commission in 1979.

After the implementation of the reform and opening-up policy, to further conduct research on input–output techniques, in June 1980, scholars including Guan Zhaozhi, Wu Wenjun and Chen Xikang of China Academy of Sciences proposed to the State Council to compile a new national input–output table and apply it to China's planned economic work. The State Council adopted the proposal and authorized the Forecast Centre of State Development Planning Commission, National Statistics Bureau and other departments to compile an input–output table of the Chinese economy in 1971. It took 2 years from the spring of 1982 to the end of 1983 to complete this table. Around this period of time, the Systemic Institute of China Academy of Sciences compiled an extended input–output table of the Chinese economy in 1979 based on the input–output table of the Chinese economy in 1973. Meanwhile, the Industrial Economic Institute of China Academy of Sciences compiled a national value-oriented input–output table of the Chinese economy in 1979 based on producer prices. Therefore, there were two input–output tables for the Chinese economy in 1979 at that time. In addition, based on the input–output table of the Chinese economy in 1981, the National Statistics Bureau also compiled an extended input–output table of Chinese economy in 1983.

In March 1987, owing to the repeated suggestion and promotion of Ma Bin and experts from China Academy of Sciences, State Council released the *Notice on Conducting National Input–Output Research*, which specified that relevant departments should conduct a national input–output analysis every 5 years and compile basic input–output tables. The institutionalization and regularization of the compilation of the Chinese input–output tables indicated that China's input–output research had entered a new stage. Thus far, the National Statistical Bureau has compiled five basic input–output tables of the Chinese economy in 1987, 1992, 1997, 2002 and 2007, respectively and four extended input–output tables of the Chinese economy in 1990, 1995, 2000, and 2005, respectively.

In March 1987, China Renmin University, the Institute of Systems Science at the Chinese Academy of Sciences and the National Statistical Bureau co-founded the China Input–Output Association. From 1987 to 1992, it was affiliated to the China Association of Quantitative Economics. In May 1993, approved by the Ministry of Civil Administration, it was registered formally as a national academic research institution affiliated to the National Statistics Bureau. After its establishment, the China Input–Output Association held an annual meeting every four years and published articles after each meeting. Since its first annual meeting held in Jiujiang, Jiangxi Province in October 1988, it had, up till now, held 7 annual meetings altogether. In 1986, Wassily Leontief, founder of input–output theory, who was interested in the development of the Chinese input–output technique, visited China and spoke highly of what Chinese input–output intellectuals had done for input–output research.

2. Compilation and Compiling Techniques of Chinese Input–Output Tables

China's research on input–output analysis grew rapidly after China's implementation of the reform and opening-up policy, which, among other things, was based on and manifested as the compilation of a large number of various input–output tables. In the process, China scored considerable achievements, accumulated rich experience and formed its own characteristics in compiling input–output tables. This can be seen from the following aspects.

2.1. *Compilation of various input–output tables*

Most countries in the world compile value input–output tables, but China is among the very few countries that have compiled physical input–output tables, which was of a great concern abroad (Bulmer-Thomas, 1982). The merits of physical input–output tables lie in their insusceptibility to price changes, accuracy in reflecting pollutants' emissions, energy production and utilization and a true reflection of economic and technical ties and structural relations as well as their changes. The first national input–output table (the table of China in 1973) is a physical one. In addition, the input–output table of China in 1981 and the input–output table of China in 1992 contain both value tables and physical tables. The physical input–output table of China in 1981 involves 146 most important physical products and the physical input–output table of China in 1992 involves 151.

Apart from national input–output tables, China's application of input–output analysis is also characterized with the compilation of sector input–output tables and corporate input–output tables. In China, the chemical industry was the first to compile input–output tables. In 1978, it compiled a physical input–output table of 16 products. Later, sectors including mechanical electronics, shipping manufacture, energy, the army and agriculture all compiled sector input–output tables. Among them, tables compiled by the Institute of Systems Science at the Chinese Academy of Sciences are especially famous for their uniqueness and have laid the foundation for the proposal of input–occupancy–output technique. As regards corporate tables, Anshan Iron and Steel compiled an input–output table 1964 based on metal balance. Afterwards, it compiled a physical input–output table 1977–1981 and value input–output tables for some certain years. Besides, at that time, Shanghai Gaoqiao Chemical Industry and other corporates also compiled corporate input–output tables which produced delightful results after being put into application.

Moreover, China also compiled regional input–output tables. The physical input–output table of 88 products and value input–output tables of 27 sectors of Shanxi Province compiled in 1979 were the first set of regional input–output tables. After 1987, except a few certain provinces, most provinces, municipalities and autonomous regions kept abreast with the state to compile annual regional input–output tables. Gradually, the compilation of regional input–output tables was standardized.

In addition, China also compiled and studied inter-regional input–output tables. Jiangsu Province once compiled an input–output model of South Jiangsu and North Jiangsu. Xinjiang once compiled an input–output model of South Xinjiang and North Xinjiang. In the late 1990s, the Development and Research Centre of the State Council, National Statistics Bureau and East Asia Development Research Centre of Kitakyushu, Japan compiled an inter-regional input–output table 1987 of seven Chinese regions and applied it to the study of interregional economic ties. In 2003, financed by the National Natural Sciences Foundation, the Development and Research Centre of the State Council and National Statistics Bureau compiled an interregional input–output table 1997 of 8 Chinese regions and developed a multi-regional CGE model based on it. At the same time, the China National Information Centre and IDE of Japan co-compiled an inter-regional input–output table 1997 of eight Chinese regions and adjusted it into an inter-regional input–output table of 8 Chinese regions 2000. Later, the China National Information Centre, along with Tsinghua University and China Renmin University, conducted an inter-regional input–output analysis of Chinese regions. In the early 2009, supported by the National Natural Sciences Foundation of China, Tsinghua University, in collaboration with China National Information Centre and National Statistical Bureau, began to compile inter-regional input–output tables of 8 Chinese regions 2002 and 2007 and started to conduct research on the dynamic input–output of Chinese regional economy.

To compare changes at different times, China compiled two sets of comparable sequence tables. One set of comparable sequence tables, tables of China 1981, 1983, 1987, 1990, 1992, 1995, was co-compiled by the National Statistical Bureau and Hong Kong Chinese University around the late 1990s and the other set of comparable sequence tables, tables of China 1987, 1992, 1997, 2002 and 2005, was compiled by the National Statistics Bureau and China Renmin University in 2008. These two sets of comparable sequence tables are important to the study of changes in long-term Chinese economic structure.

Besides, in the early days of the development of Chinese input–output analysis after China's implementation of the reform and opening-up policy, He Keng and other Chinese scholars also compiled an information input–output table of Yueyang, Hunan Province and a national information input–output table. The Statistics Bureau of Shanxi Province also compiled labor consumption input–output tables.

2.2. *The establishment and improvement of input–output compilation system and new System of National Accounts*

In 1968, the System of National Accounts (SNA) of the United Nations included input–output tables into its accounts system for the first time. Thus, input–output tables, GDP tables, capital flow tables, international balance tables and balance sheets became the five basic account tables of China's system of national accounts. From then on, the relationship between input–output and national accounts becomes closer.

The compilation of input–output tables promoted the establishment and development of a new Chinese system of national accounts. When we compiled the first input–output table 1987, instead of copying the input–output table of physical products of MPS, we drew upon the strength of SNA and made an input–output table that included both physical production activities and non-physical production activities, which contributed to our compilation of SNA-style input–output tables (Qi, 2003). When compiling the input–output table 1987, we invented a "plate-style" transformative structure and achieved the alternation of MPS and SNA through the adjustment and auxiliary computation of plates (Lin, 1988). In so doing, we not only conformed to the Chinese reality, but also took into account the connection of two accounts systems. By the time we compiled the input–output table 1992, we happened to experience the transition from MPS to SNA. Although we still conducted basic researches within MPS framework, we compiled the input–output table 1992 "according to the theory of new Chinese system of national accounts" in terms of index specification and overall design and therefore "it was compatible with our new system of national accounts" (Qi, 1995).

In terms of the structure of input–output tables, the intermediate quadrant shows the intermediate input and intermediate utilization of product sectors, the final product quadrant shows the composition of final demands and the initial input reflects the income allocation. The three quadrants correspond with the GDP aggregate from different angles and are interrelated. Such kind of inter-relationships, on the one hand, provides a mutual confirmation of GDP auditing and on the other hand, reflects the relationship between medium and macro, thus offering a micro-interpretation of a macro-concept. However, our current practice is different from international practice. In our country, we usually use GDP aggregate data as the controller, which makes it difficult for us to use input–output techniques to test and confirm the accuracy of data. Therefore, the improvement of the compilation system and methods has become an integral part of the development of our auditing system.

2.3. *Improvement of compiling methods*

When it comes to the improvement of compiling methods, China not only draws upon the experience of foreign countries, but also takes into consideration the Chinese reality and has developed its own characteristics.

The major reason why we compiled our first national input–output 1973 as a physical one is that under the planned economic system, data on physical products are easier to collect and acquire. In addition, it is also because most national goals are measured by physical products, which makes it easier to formulate and assess national plans (Chen, 1989). This shows that when selecting input–output tables, we pay attention to the Chinese reality.

On the whole, compilation methods can be divided into two categories: direct decomposition and indirect deduction. Indirect deduction involves the use of supply tables and utility tables to deduct a symmetrical input–output table while direct decomposition involves the use of auditing data and input–output research to compile tables. On the whole, we tend to use direct decomposition when compiling tables, except in some particular years when we use indirect deduction. For instance, we mainly used indirect deduction to compile the national input–output table of 1981. When summarizing our experience in compiling the input–output table of 1981, we stated that "decomposition and deduction as two methods are not exclusive or contradictory. On the contrary, they are complementary and can be used together" (Mao & Mao, 1988). Since then, we usually adopt direct decomposition as the major method and indirect deduction as a complementary when compiling national input–output tables. Medium- and large-sized enterprises usually prefer direct decomposition while small enterprises whose statistic system is not advanced enough tend to use indirect deduction.

In short, when compiling input–output tables, China has always respected its own conditions and realities and given prominence to the quality of data of the tables. With the integration of more input–output tables into the system of national accounts, changes in both foreign and domestic accounts systems and shifts in international compilation ideas, we need to update our compilation thoughts to meet the new requirements.

3. Progress in Theoretical and Applied Input–Output Research

Over the past three decades since China's implementation of the reform and opening-up policy, China has made constant achievements in not only the practical compilation of input–output tables, but also theoretical input–output researches.

The most important achievement in theoretical input–output research was scored by Chen Xikang. In 1989, he became the first to put forward the notion of input–occupancy–output technique at an international conference. According to Chen, input–occupancy–output technique not only studies the relationship between the input and output of products in different sectors, but also approaches the relationship between the output of an individual sector and its fixed assets, labor force and natural resources. The input–occupancy–output technique attracted international attention and was well-acclaimed. For instance, W. Isard, Director of America Academy of Sciences regarded it as "a fairly valuable discovery" and "ground-breaking research"; Wassily Leontief, Nobel Prize Winner in Economics remarked that "the calculation method of input–occupancy–output and complete consumption coefficient is a very important invention and innovation in the field of input–output analysis".

Apart from Chen Xikang, other scholars have also made great achievements in theoretical input–output research. For instance, Liu Qiyun proposed to develop an input–output model based on distribution coefficient and establish a complete analytical system of symmetrical models; he also proposed to develop a two-stage input–output model to structuralize the Keynesian Multiplier. Zhang Shouyi proposed to add a calculation model of price changes in a sector to the accumulation and consumption part, develop a technology input–output model to measure the contribution of technological progress to economic growth and develop a labor–consumption model to study prices. To deal with the inconsistency between two formulas in the mixed handicrafts hypothesis of the *System of National Accounts* of the Statistics Bureau of the United Nations, He Juhuang put forward a complementary hypothesis and mathematical deduction and elicited three major formulas. Gong Xiaoning explored the relationship between final demands and income distribution within the input–output framework.

In terms of practical research, based on theoretical research, quantitative research that made full use of the advantages of input–output technique was conducted to cope with the major and prominent practical economic problems that appeared at different periods of time to offer advice to economic policy-making.

After a long-term research, Chen Xikang proposed to use the synthetic factor forecast method to forecast national crop output, whose major techniques is input–occupancy–output technique. During the 29 years from 1980 to 2008, this method, with an average error of 1.9%, remained unparalleled by other methods in terms of forecast technique and application effects. The forecast project progressed smoothly and was well-acclaimed by the central government. It was the first prize winner at the International Operations Research Progress Awards, Beijing Science and Technology Awards, and Scientific and Technological Progress Awards of the Chinese Academy of Sciences and winner of the first Management Excellence Award and the first Outstanding Achievement in Science and Technology Award of the Chinese Academy of Sciences.

In the key research of the National Natural Sciences Foundation on nonlinear and dynamic input–occupancy–output techniques and its application led by Chen Xikang, we made progress in the nonlinear important coefficients input–occupancy–output technique, computable nonlinear dynamic input–output model, and new dynamic input–occupancy–output models that take into consideration human capital and technology. For instance, we built a nonlinear input–output model with a technological progress function as the important coefficient whose accuracy was higher than that of RAS by an average of 6.59%; by unifying Leontief's linear dynamic model and computable nonlinear dynamic general equilibrium model, we obtained a computable nonlinear dynamic input–output model and a better way of calculating balanced growth orbit and optimal growth orbit; we formulated a new dynamic education-economy input–occupancy–output model and an education-economy input–occupancy–output model that takes multiple time lags into consideration.

In 2007, Liu Qiyun from China Renmin University directed the key research project of National Social Science Fund on "strategic adjustment of structure and transition of growth style of the Chinese economy". Adopting the theoretical framework of multi-sector input–output model and relying mostly on quantitative analysis, he conducted a simulation of real changes, internal economic ties and policy effects of the Chinese economy amidst its structural transition and derived some concrete and feasible polices and suggestions. These policies and suggestions were of great reference value to the successful transition of the Chinese economic growth style. The entire research project lasted for 3 years and was not completed until the end of 2009. In addition, the Input–Output Research Group of China Renmin University participated in the research on the Macro-Economic Report published regularly by China Renmin University. It applied input–output techniques to the analysis of major economic problems in China at different times and put forward some suggestions, which were of enormous social influences.

Between 2005 and 2006, China Renmin University applied input–output technique to the analysis of the economic influences of water, coal and petroleum price changes. Their research result was adopted by Beijing Commission of Development and Reform as a reference when studying and fixing prices of resources and products.

The Institute of Systems Science at Chinese Academy of Sciences, in collaboration with the National Statistics Bureau and China Renmin University, compiled a water resources input–occupancy–output table for China and water resources input–occupancy–output tables of nine major river basins in China to calculate water resources investment benefit, the shadow price of water, the optimal percentage of water investment in GDP and in national fiscal expenditure. This research project passed the test organized by the Institute of Water Resources in 2002 and was spoken highly of by evaluation experts.

Researchers from the Chinese Academy of Sciences used the local block input–output model and econometric method to measure the effects of the standard for energy efficiency of public buildings on economy and environment. The policy proposed by the researchers was approved by Chinese Vice-President Li Keqiang. When conducting the research project, "Beijing High-Tech Industry–Traditional Industry–Labor Force Input–Occupancy–Output Model and its Application", people compiled a Beijing High-Tech Industry–Traditional Industry–Employment Input–Occupancy–Output table which revealed the interaction between the development of high-tech industry and labor force demand and illustrated the relationship between high-tech industry and employment. The Beijing Development and Planning Commission spoke highly of this project, the research findings of which were adopted and applied to the Tenth Five-Year Plan of Beijing National Economic and Social Development and the Eleventh Five-Year Plan and the project was awarded the Second Prize at the Beijing Scientific and Technological Progress Awards, 1998. Scholars represented by Guo Ju'e from Xi'an Jiaotong University built a theoretically advanced and practically workable financial input–occupancy–output model. They also compiled financial input–occupancy–output tables of China 1997, 2000 and 2001 and put them into application. Their efforts were highly recognized by the society. Besides, in the recent research project, "A study on residential consumption potential in different industries against the background of financial crisis", scholars represented by Liu Xiuli used input–output analysis and the quantitative economic method, and conducted model analysis and measurement and proposed suggestions for coping with financial crises and further expanding residential consumption. Based on their research, they wrote a report which was submitted to the Central Government and regarded as "an important report" by Chinese Vice-President Li Keqiang, who hoped that researchers would "come up with new measures through their studies that can boost industrial development and promote employment". The National Development and Reform Commission and experts from the National Information Centre paid a special visit to the Institute of Systems Science at the Chinese Academy of Social Sciences to study this report and discuss the implementation of the polices and measures proposed in the report. In the 21st century, with the compilation of inter-regional input–output tables of China, researches on inter-regional input–output developed rapidly. The Development and Research Center of the State Council built a CGE model and conducted a multi-regional economic analysis of China based on its Inter-Regional Input–Output table of 8 Chinese Regions 1997. Tsinghua University, the National Information Centre and Institute of Developing Economies used their Inter-Regional Input–Output Table of 8 Chinese Regions 2000 to conduct a multi-dimensional and comprehensive analysis of Chinese regional economic development. Their research findings also attracted consideration attention both at home and abroad.

4. International Exchanges

At the same time, when domestic researches on input–output were enjoying tremendous growth, international exchanges and cooperation also developed rapidly. The National Statistics Bureau and Norwegian Statistic Bureau made a joint effort to conduct the "China–Norway Environmental Statistic and Analysis Project", compiled energy accounts, made an air pollutant emission inventory and built an integrated analytical and predictive model of China's environment and economy. This research project provided basic data for the research on China's energy management and sustainable development. Moreover, since it also built models to conduct a simulation of economic policies, it also provided a basis for formulating energy policies and determining environmental protection measures. The Institute of Mathematics and Systematic Science of the Chinese Academy of Sciences, along with Chinese University of Hong Kong, Hong Kong University of Science and Technology and the University of California, conducted research on the stimulating influence of Sino–US bilateral trade; on the development the Chinese and the American economies. The research result showed that in terms of the stimulating influence, the actual surplus of Sino–US trade was much smaller than what was published. The research result attracted great attention of related government departments and the intellectual world. The research report based on it, provided useful information to President Hu Jingtao who visited the United States in April 2006, played an important role in Sino–US trade negotiation and was of a great concern to both Chinese and American senior officials. The United States International Trade Commission (USITC) and scholars from the European Union (EU) attached great importance to this method and offered to cooperate with China. In 2008, the Institute of Mathematics and Systematic Science of the Chinese Academy of

Sciences, USITC and University of Groningen made a concerted effort to apply for the International Cooperation Major Project of Natural Sciences Fund Commission (NSFC). Their application was approved and their research result was awarded Sun Zhifang Economic and Scientific Articles Award in 2008 and the Second Zhang Peigang Developmental Economics Award.

Between 1998 and 2004, sponsored by Alliance for Global Sustainability and National Science Foundation, United States (NSF, US), the Institute of Mathematics and Systematic Science of the Chinese Academy of Sciences, in collaboration with Tokyo University and other universities, compiled input–occupancy–output tables of Chinese township enterprises and carried out research on the energy utilization and environmental protection of them. The United Nations Industrial Development Organization also authorized Yang Cuihong and Chen Xikang to conduct research on the coal refinement industry of Shanxi Province. The project group reported their research findings many times at various conferences and in related institutes including the Massachusetts Institute of Technology, University of Tokyo, the Annual Meeting of AGS held in Costa Rica, and the Annual Meeting of Coal Industry held in North America, which attracted considerable attention of conventioneers.

In addition, the National Information Centre and Institute of Developing Economies, Japan has cooperated with each other since 1988 and participated in the compilation of input–output tables of Asian countries (regions) including China.

Apart from cooperation in research projects, international and personnel exchanges were conducted on a regular basis. Since 1986, the Institute of Systems Science at the Chinese Academy of Sciences maintained a long-term cooperative relationship with Professor Karen R. Polenske from Massachusetts Institute of Technology in energy utilization, environmental protection, utilization of water resources, research on regional input–output models, etc. The Institute of Systems Science at the Chinese Academy of Sciences and the Faculty of Management of School of Graduates at the Chinese Academy of Sciences also cooperated extensively with University of Groningen through holding regular symposia, sending visiting scholars to each other and establishing dual Ph.D. programs, etc. In October 2006 and October 2008, the China–Holland Input–Output Seminar was held in Groningen and Beijing respectively. A number of prestigious input–output experts from China, the Netherlands, the United States and Japan attended the seminar. The Chinese Academy of Sciences and University of Groningen Dutch established and sponsored a dual-Ph.D. program: KNAW-CAS and sent visiting scholars to both sides. So far, 8 Ph.D. students at the China Academy of Sciences have been selected to participate in the program and are doing a Ph.D. degree under the instruction of their supervisors. In addition, the Chinese Academy of Sciences also established a cooperative relationship with Illinois University, US. From April 2008 to January 2009, Liu Xiuli from the Chinese Academy of Sciences paid a visit to the Applied Laboratory of Regional Economy at Illinois University, US and cooperated with the President of the Association of International Regional Economy and Professor Geoffery Hewings, a prestigious input–output scholar, on Sino–U.S. energy consumption and greenhouse gas emission.

China's input–output research has also aroused great attention of the international community. In June 2005, the 15th International Input–Output Technique Conference was held at China Renmin University, Beijing by the International Input–Output Association, Chinese Input–Output Association, China Renmin University, National Statistics Bureau and Development and Research Centre of the State Council. Professors, researchers and government officials from over 40 countries and regions attended this conference.

5. Future Development Prospects

Since input–output analysis was introduced into China, it has achieved a great deal, but its development has also met some setbacks. In particular, since the establishment of the goal of market reform, due to a misunderstanding of input–output methods, some scholars believed that input–output was related to the planned economy and therefore, had gone out of date. In fact, as a quantitative economic method, the input–output method has its own characteristics like other quantitative analytical methods and one of its major characteristics is systematicness. Liu (2006) summarized the systematicness reflected by the input–output method in three major points:

First, systematicness reflected economic ties. The three quadrants in an input–output table are closely related and this relationship is exactly an objective description of economic ties. Within an input–output framework, the quadrant of intermediate product and coefficient of direct consumption reflects the technical ties in the production process, the quadrant of final product reflects the final demand structure and the third quadrant reflects the income distribution relationship. In other words, the three quadrants are systematically related. Therefore, in an input–output framework, we can analyze production and growth according to row-based coefficients or analyze price and income distribution according to column-based coefficients and establish a correlation between growth and income distribution. We can also study how final demand and income distribution can influence technical selection, etc.

Second, systematicness is reflected by the combination of structural analysis and aggregate analysis. An economic process involves a structural relationship and an aggregate relationship. Both structural analysis and aggregate analysis have their own limitations, and it is through input–output method that they are combined.

Three, symmetry is reflected by input–output method. For instance, Wassily Leontief built a standard input–output model based on a column-oriented coefficient. In addition, we can also build models based on a distribution coefficient based on symmetry, flow relations, inventory relations, etc.

Another characteristic of input–output analysis is integrity. Today, we have already noticed that the input–output method as an experimental research method has produced numerous results based on an extension of basic models in structural analysis, regional study, price analysis, multipliers analysis, income distribution study, resources and environmental issues study, computable general equilibrium, etc. As an accounts system, input–output method has become an integral part of the national accounts system.

Besides, input–output is also an economic theory in its own right. In their seminal work *Linear Programming and Economic Analysis*, Dorfman, *et al.* (1958) considered input–output, linear programming and game theory as three dominant methods in linear economics. In neoclassical analytical systems, on the other hand, the input–output model was considered as a special productive function and an exception of the general equilibrium model whose technique was replaceable (Koopmans, 1951). On the other hand, Leontief system has represented the classical analytical tradition since the era of Francois Quesnay and laid the foundation for the combination of the theory and practice of the Sraffa–Leontief System. The research on input–output system as part of theoretical economics not only improves the method itself, but also promotes our understanding of Western input–output theory. It is of great significance to both the innovation of economic research methods and the application of economic methods in our analysis of practical Chinese economic problems.

Given that input–output is a complete system, to command input–output and apply it to our analysis of actual problems, we need to understand not only the method itself, but also data involved in input–output theory. "In an input–output model, every letter represents a specific economic index, every equation indicates a kind of economic quantitative relation and every mathematic deduction reflects a change in economic ties. Input–output is not a hollow economic theory or merely a mathematical method. Instead, it is an exemplary integration of economic theory and mathematical method" (Liu *et al.*, 2006). Obviously, efforts in this field will bring new success to our input–output research in a new era.

Bibliography

Bulmer-Thomas, V. (1982). *Input–Output Analysis in Developing Countries*. Chichester: John Wiley & Sons Ltd.

Chen, X. (1989). Input–output techniques in China, *Economic Systems Research*, 1(1).

Dorfman, R., Samuelson, P.A. & Solow, R.W. (1958). *Linear Programming and Economic Analysis*. New York: Dover Publications, Inc.

Koopmans, T.C. (1951). *Activity Analysis of Production and Allocation*. New York: John Wiley & Sons, Inc.

Lin, Xianyu (1988). Conversion Method of the Two Accounting Systems of the National Input–Output Table, in *Contemporary China's Input–Output Theory and Practice*, edited by Chen, Xikang. Beijing: China International Broadcasting Publishing House.

Liu, Q., *et al.* (2006). *Input–Output Analysis*. Beijing: China Renmin University Press.

Mao, Bangji & Mao, Zurong (1988). The Compilation and Application of the 1981 National Input–Output Table, in *Contemporary China's Input–Output Theory and Practice*, edited by Chen, Xikang. Beijing: China International Broadcasting Publishing House.

Qi, S. (1995). *The Characteristics, Techniques and Methods of National Input–Output Survey 1992 and Input–Output Table of China 1992*. Beijing: China Statistics Publishing House.

Qi, S. (2003). *An Introduction to the Compilation of Chinese Input–Output Tables and Their Application*. China Statistics and Information Network.

Zhong Qifu (2007). *Selected Readings of Zhong Qifu*. Beijing: China Renmin University Press.

Three Developmental Stages and Tasks
of China's Econometrics*

Li Zinai[†]

1. Introduction

Over the past three decades since China's implementation of the reform and opening-up policy, China's econometrics has undergone profound changes. Among them, the development of econometrics is the most significant one. This can be seen from the empirical data below. According to our research, the proportion of finance and economics faculties (or economics as it has been called since 1992) at institutions of higher learning offering "econometrics" was 0% in 1980, 18% in 1987, 51% in 1993, 92% in 1997 and 98% in 2006. The data may allow for some errors due to variations in research scope and sampling methods, but on the whole it reflects the accessibility of econometrics in institutions of higher learning at different times. Nowadays, econometrics has become a core course of college students studying economics.

Based on our statistical analysis of over 3,100 papers published in *Economic Study* from 1984 to 2006, the percentage of papers taking econometric models as major analytic methods in all papers was 0% in 1984, 5% in 1992 and 11% in 1998. The figure then soared to 40% in 2004, 56% in 2005 and 53% in 2006. In addition, its research objects came from all economic fields and its models were applied to all its branches. This is also the same with other economic journals such as *Financial Study* and *World Economy*. There is no doubt that econometric models have

---

*This chapter came from the 11th issue of *Dynamics in Economics*.
[†]Li Zinai, Professor of the School of Economic Management of Tsinghua University and Doctoral Supervisor.

become a major empirical research method of theoretical economic study and practical economic analysis in China.

From our knowledge, the number of doctoral programs in quantitative economics (with econometrics at its core) in China rose from 1 in 1984 to 2 in 1993, 7 in 1998, 18 in 2004 and 25 in 2006 (paramount to doctoral bases of finance and outnumber other disciplines of economics). If we count the number of econometric programs offered by other disciplines of economics such as management sciences and engineering and other secondary disciplines of applied economics, the figure will be doubled. Thus, we can see that the relatively new discipline of econometrics has developed faster than other disciplines of economics over the past three decades.

2. Factors of the Rapid Development of Econometrics in China

What makes Chinese econometrics develop so rapidly and promote it to such an important position? The factors are in Sections 2.1–2.5.

2.1. *Econometric theory is scientific and econometric models are essential to empirical economic research*

All scientific research, whether they are natural sciences or social sciences, or even human sciences, must follow the procedures below:

Firstly, observe contingent, individual and special phenomena.

Secondly, propose hypotheses, theories or models that apply to the certainty, generality and universality of phenomena based on the prior observation.

Thirdly, test hypotheses with experimental methods, prediction methods, regression methods and other widely-used methods.

Lastly, summarize the law of certainty, generality and universality. In economic research, if we advance a hypothesis (a theory or a model) purely based on our observation and test it without imposing on it any value judgment, we are doing what is called an empirical research.

Empirical research includes both theoretical and experimental empirical research. When a piece of economic research is going through a test, experimental empirical research is scientific and easy to conduct. However, economic issues are beyond experimentation. In other words,

regional macro-economic models came into vogue. In particular, the Annual Economic Analysis and Prediction Conference held by the China Academy of Social Sciences since 1990 that releases reports on economic analysis and prediction and publishes the "Chinese Economic Analysis and Forecast" *Bluebook*, tremendously promoted the development of research on applied econometric models, and is influential both at home and abroad.

In November 1993, the Third Plenary Session of the Fourteenth Central Committee of CCP set the goal of Chinese economic reform to develop a socialist market economy. This, together with the trend of global informatization, had profound influence on the teaching of economic specialties at institutions of higher learning and promoted the popularity of econometric courses.

Between 1994 and 1995, support by the former Department of Higher Education of the State Education Commission, Chinese Association of Quantitative Economics and the Special Committee at Institutions of Higher Learning of Chinese Association of Quantitative Economics mobilized over 20 institutions of higher learning to compile *Econometrics Teaching Program* which included *Econometrics Teaching Program (Finance and Economics)* and *Econometrics Teaching Program (Integrated, Science and Engineering)*. It served as a useful guide to econometrics teaching. According to a March 1997 research study, of all institutions of higher learning with economic specialties, 92% offered econometrics and 88% offered econometrics as a compulsory course with an average of 54 class hours.

In 1996, former State Education Commission launched *Plan for Reforming Teaching Profile and Curriculum System of Advanced Education in the 21st Century*. In July 1996, it designated "Study on Curriculum System and Teaching Profile of Quantitative Analysis Courses of Economic Management" as one of the 20 major research projects of economic management. Based on a survey and analysis of the *status quo*, demand for quantitative economic specialties and curriculum of economic specialties at universities of developed countries, the former State Education Commission proposed in September 1997 to set up three levels of quantitative analysis courses for economic undergraduates: Core courses, compulsory courses and selective courses and include econometrics into the core courses.

In July 1998, at the First Conference of the Teaching Supervision Committee of Ministry of Education of Economic Disciplines and Specialties at Institutions of Higher Learning, econometrics was designated as a common core course for all economic students at institutions of higher learning. This event symbolized the modernization and scientization of Chinese economic teaching. It influenced the production of Chinese economic talents and indicated that western econometrics had been popularized in China.

3.2. *The stage of improvement and widespread application*

The second stage (from 1998 to 2010) was known as the improvement of econometrics teaching and widespread application of econometrics. It began with the determination on econometrics as a core course and ended in July 2006 with the first International Academic Meeting held by the World Econometrics Association in Mainland China. The popularization of undergraduate econometrics courses created an opportunity for institutions of higher learning to offer postgraduate econometrics courses.

In the late 1990s, with the introduction of a number of advanced econometrics textbooks into China and the publication of advanced econometrics textbooks compiled by Chinese scholars, institutions of higher learning began to offer advanced econometrics courses to economic postgraduates in the real sense.

Undoubtedly, that the Nobel Prize in Economics 2000 and 2003 was awarded to Daniel Kahneman and Clive W.J. Granger, econometricians who had made extraordinary contributions to microeconometrics, modern macroeconometrics and financial econometrics, enormously propelled the development of advanced econometrics teaching. According to the National Seminar on Producing Quantitative Economic Ph.D. students held in October 2006, all institutions of higher learning put advanced econometrics at a prominent position of all Ph.D. courses. Some schools offered more than 1 advanced econometric course and some offered as many as 6 advanced econometric courses. The opening of advanced postgraduate econometric courses also improved the curriculum system and teaching standard of undergraduate econometrics courses. In particular, marked progress was made in the integration of basic and frontier econometric courses and the combination of econometric theory and practice. A number of quality courses at national and provincial levels were set up.

According to a survey in 2006, 98% of institutions of higher learning offered econometric courses to economic students. Among them, 69% offered all econometrics courses as compulsory courses, 36% offered some as compulsory courses and 3% offered all as selective courses. The data was close to those in the survey of 1997. However, curriculum contents had experienced remarkable progress. According to the survey, 100% of institutions of higher learning introduced classical mono-equation linear models, 73% introduced classical simultaneous equation model, 40% introduced prevalent applied models, 26% introduced time-series analytical models and 15% introduced micro econometric models.

At the same time, econometric models were extensively used in our economic theoretical research and economic problems analysis. As afore-mentioned, the percentage of articles that adopted econometric models as major analytic tools in all articles published in economic academic journals rose and exceeded 50%. This percentage was similar to that of econometric articles in *American Economic Review* and other journals of its kind.

In addition, its research objects were extended to all fields and its models were applied to all branches of econometrics. The econometric model has become a mainstream empirical research method in our economic research. More important is that it has become a common tool of integrated economic management departments and economic research institutes to analyze economic situations, study practical economic problems and formulate economic policies, and has improved economic prediction and decision-making.

In July 2006, the world econometric association held an international academic meeting on the Chinese Mainland for the first time. Nearly 500 scholars from 37 countries and regions participated in this meeting and 50 Chinese scholars had their articles passed the standard test and were selected to be read at the meeting. In 1987, I attended the international academic meeting of the World Econometric Association held in Tokyo which was seen by the then president of world econometric association as the first time that Chinese scholars had attended the international academic meeting of world econometric association. In two decades from then on, this leading international academic meeting was held in the Chinese Mainland. This manifested the improvement of our teaching in econometrics and the development of our applicable research.

3.3. *The stage of development of and innovation*

The third stage which began in the recent 2 years can be seen as development and innovation of our teaching and research in econometrics. At this stage, our major tasks include: Strengthening our theoretical research, improving our applicable research and developing our own courses in econometrics.

4. Three Major Tasks Facing Our Teaching of Econometrics at Present

4.1. *Strengthen the research and innovation of theoretical methods of econometrics*

The evolution of econometric theoretical methods from classical methods to modern methods has formed a widespread system. In the first two decades of our econometric development, Chinese scholars, rather few in number, have always been studying and following the practices of foreign scholars. Even among postgraduates in econometrics, less than 10% chose econometric model as their research topic. Therefore, we are far behind by the international community in this regard. In recent years, with the return of some distinguished overseas econometric scholars and well-trained young overseas scholars in economics as well as the vigorous international academic exchanges, the situation is undergoing some changes. Theoretical research of econometrics not only underpins the development of econometrics, but also manifests how developed our econometrics is. Only by strengthening theoretical research and producing original research results can we make headway toward the mainstream of world econometrics. Therefore, it is one of our major tasks to strengthen theoretic research in econometrics.

The problem is how to strengthen our theoretic research in econometrics. When we unfold an authoritative theoretical journal in econometrics in reverence, we cannot help asking the following questions: What is creative research in theoretical methods of econometrics and what is the orientation of our research and innovation in theoretical methods of econometrics? A review on the history of econometrics revealed to us that econometrics developed after problems. In other words, once new problems emerged, theoretical methods that dealt with them would develop accordingly. This is particularly true with the major developments in modern econometric theories. Based on his discovery of the selectivity of sample data and its influence on classical econometric model, James Heckman developed the selective sample model method; by studying the relationship between selective results and workable factors, Daniel L. McFadden developed the discrete selective model; discovering the high probability of pseudo regression between non-stationary time sequences. Clive W.J. Granger advanced the method of testing the stability of time sequences and established a structural model between non-stationary time sequences. These major innovative findings that have claimed Nobel Prize in Economics all come into being following the emergence of economic problems. Thanks to the sustainable and steady development of the Chinese economy, research on the Chinese economy has become an integral part of international mainstream economics. At the same time, however, the particularity of Chinese economic development and data has also presented many new challenges to the innovation of theoretical econometric methods, such as structural change characteristics of economic development in time series, the enormous difference of individual behaviors at the same time section and the irrationality of selective behaviors. To solve these problems and develop theories and methodology in the process should be the key direction of our theoretical research in econometrics.

4.2. *Improve applied research in econometrics*

As applied research in econometrics is in full swing, it is imperative that we improve the quality of our applied research. At present, misuse and abuse of econometric models is a recurrent or even common phenomenon. While the former leads to a kind of "self-deception" among Chinese econometricians, the latter leads to their "personal amusement". Under such circumstances, the question of how to enhance the standard of applicable research in econometrics deserves our serious consideration. Otherwise, public enthusiasm and confidence in econometrics will be impaired and the bright future of econometrics may be well ruined. Reasons for the appearance of these phenomena are manifold. First and foremost, they occurred due to our lack of profound study and accurate understanding of methodological basics of econometric models including: Philosophical basics, economic basics, mathematical basics and statistical basics. Problems include: Scientization of econometric modeling, reliance of models on data, economic relations-oriented general model specification, relativity of model variation specification, originality of model random disturbance term, asymmetry of hypothesis test, limitations in model application, etc. (Li Zinai, 2007; Hong Yongmiao, 2007). Therefore, it is important and urgent to conduct research on methodological basics of econometric model methodologies to improve our applicable research. To fulfill this task, the most important thing is to set a proper overall model (Li Zinai, 2008).

It should be the guiding principle of our applied econometric research to study major problems, adopt proper models and aim at important findings. The importance of studying major problems is obvious. The crux is: What are the major problems? Certainly, we should pay attention to and study those problems favored by Western economic journals. However, we should pay even greater attention to major practical problems facing our own reform and development. To adopt proper models and methods is to develop our own models out of models designed by foreign countries based on the characteristics of problems and data that we are approaching. Aiming at important findings is using econometric modeling to discover those laws that cannot be detected through general observation and qualitative analysis. If the results of our research based on econometrical models and complicated estimation and tests are the same as what people have been well familiar with, the value of our research will decrease.

We should pay special attention to the development of microeconomics in the social field. According to our statistical analysis of economic journals, the existing applicable research in econometrics concentrates on macroeconomics and financial markets, each accounting for over 30%. There are two reasons: First, as the greatest concerns in our economic development, the macro economy and financial markets merit our serious study; second, they are easy to study as data on the macroeconomy and financial markets are easy to acquire. The Seventeenth Congress of CCP proposed the scientific outlook on development as our guideline and set the goal of building a harmonious society. Since the key to building a harmonious society is to improve people's livelihood, problems related to

A Review on the Development of Economic Prediction in China*

Wu Jiapei and Liang Youcai[†]

Economic prediction, structural analysis and policy analysis are 3 applied fields of quantitative economics. In view of the 30th anniversary of the founding of Chinese Association of Quantitative Economics, let us have a review of the development of economic prediction in China.

1. Emergence of Economic Prediction

From China's implementation of its reform and opening-up policy in 1978, the Chinese economy began to transit from a highly centralized planned economy to one characterized with a planned commodity economy and a socialist market economy. Before the transition, the market economy along with its corresponding economic prediction was seen as the product of capitalist society. At that time, China's economy was a completely planned one without any prediction. In June 1982, Wu Jiapei proposed in an article that China should develop its own economic prediction mechanism. The article made a comparison between economic planning and economic prediction and clarified their relationship. It also pointed out that the development of economic prediction would contribute to the economic plan and decision-making. In November 1982, the then National Planning Committee took the lead to establish an economic prediction center which conducted both macro-economic prediction and international

*A feature report written for the 30th anniversary of the founding of the Chinese Association of Quantitative Economics.

[†]Liang Youcai, Researcher of National Information Center and Doctoral Supervisor.

exchanges and cooperation in economic prediction. In January 1987, the National Economic Information Center was established, which was later renamed The National Information Center, a name given by Comrade Deng Xiaoping. When it was established, it split the former Economic Prediction Center of State Planning Commission into the Economic Prediction Ministry and the Economic Information Ministry, which continued to promote China's economic prediction. The Economic Information Ministry deals with the analysis of economic situation; it monitors national economic data and indexes time sequences. The Economic Prediction Ministry conducts short-run and medium- or long-run macro-economic predictions. It forecasts the development of the national economy with all kinds of economic mathematical models. The national information center also carries out industrial prediction, such as prediction and analysis of industrial development. Besides this, the national information center also unites the economic prediction institutions of all provinces and municipalities through its economic information system and forms a national economic prediction network.

In 1985, to expand foreign exchanges and learn advanced economic prediction methods from foreign countries, the State Council allowed scholars to attend the World Economic Prediction Project Link. This was set up and led by Laurence Klein, foreign advisor of National Information Centre and a Nobel Economics Prize winner in 1980. Later, led by Wu Jiapei, Fan Murong and Zheng Shaolian, related departments in China began to study the Chinese macroeconomic model that can be applied to Project Link. In 1986, the State Planning Commission sent Liu Youcai to the Link Center at the University of Pennsylvania, USA for a period of research and study. Under the instruction of Lawrence Klein, Liang Youcai designed a Chinese macroeconomic model. In December 1986, Liang's Chinese macroeconomic model was officially integrated to Project Link and replaced the previous Chinese macroeconomic model designed by Lawrence Liu from Stanford University, USA. From then on, the Chinese Link Project Team provided Chinese economic predictions to the Link Centre twice every year and was responsible for updating the Chinese macroeconomic model in Project Link. Dr. Hong Pingfan, director of United Nations Global Economic Monitor Department from the China National Information Centre who is in charge of daily work of the Link

Project now, has made great contribution to the design and update of the Chinese macroeconomic model.

In the late 1980s, with the popularization of economic mathematical models and the development of data and information collection, the number of departments engaged in national macroeconomic predictions rose. Apart from the China National Information Center, 710 departments including the Prediction Department of Development and Research Center of State Council, the Institute of Quantitative Economics and Technology at the Chinese Academy of Social Sciences, the Institute of Systems Science at the Chinese Academy of Social Sciences and the Department of Aeronautics and Astronautics also conducted macroeconomic prediction. Besides, sector prediction like foreign trade prediction, regional prediction like Shanghai economic prediction, enterprise prediction like input–output analysis and prediction of Anshan Iron and Steel Company as well as market and product prediction also developed.

At the beginning of September 1988, the First National Economic Prediction Conference was held in Beijing. In terms of the unprecedented presence of this conference, economic prediction has been the primary work of economic planning, which is an important part of economic management in China. It makes economic planning more scientific and helps reduce the uncertainties in economic development along with its impacts.

Since the implementation of the Sixth Five-Year Plan, and especially during the Seventh Five-Year Plan, China's economic prediction work has progressed smoothly. This could be seen from the regularization of monthly economic monitoring and quarterly economic prediction, gradual standardization of semi-annual regional economic prediction, great accuracy in annual economic prediction as well as the full development of market prediction. More than 20 provinces and regions had conducted market prediction on 16 kinds of products in 10 categories.

The number of economic prediction institutions and teams is on the rise. They have employed and developed various economic prediction techniques, simple or complicated and devised various economic prediction methods, including Delphi investigation, time series, index analysis, element analysis, etc. In element analysis alone, many models such as input–output model, econometric model, linear programming model and system dynamics model are used.

While the practice of economic prediction is making progress, remarkable progress is also being made in the research on economic prediction theory and methods. The Chinese Association of Quantitative Economics organized the publication of a series of economic research achievements, such as *Economic Prediction Symposium* (edited by Li Changming, etc. and published in 1986 by Liaoning People's Publishing House) and *100 Cases of Economic Prediction* (edited by Li Changming, etc. and published in 1986 by Liaoning People's Publishing House). It also entrusted Professor Feng Wenquan from Wuhan University with the compilation of *Principles and Methods of Economic Prediction* (Wuhan University Press, 1986). This book was in the same train of thought as other books such as *Economic Prediction and Decision-Making Techniques* (Wuhan University Press, 1983) and *Economic Prediction and Economic Decision-Making Techniques* (Wuhan University Press, 1990; 1994; 2002; 2008), this book was published a total of six times, making it the most influential one. The prefaces for these economic prediction books were all written by Wu Jiapei. Each preface points out that each individual book is different from the others, but that they are part of a continuum.

To commemorate the contribution Professor Lawrence Klein made to economic prediction with econometric models, Professor M. Dutta from University of Rochester, USA edited and published *Economics, Econometrics and the Link* (North-Holland, 1995) which collected articles written by scholars from many countries and regions. It also included the article "China's Macroeconometric Model for Project Link" written by Chinese scholars Wu Jiapei, Liu Youcai and Zhang Yaxiong.

2. Refinement of Economic Prediction

Since the 1990s, China's economic prediction became more institutionalized and standardized. This can be seen from activities and achievements of some major economic prediction departments.

Firstly, the Prediction Department of State Information Center gave a copy of internal national (regional) economic prediction report of the next year to all conventioneers of the National Planning Conference (later called National Reform and Development Conference). From 1992, it compiled the grand annual report on economic prediction: China's Economic Prospect and issued economic prediction information to the public, which attracted enormous social attention. Former Vice-Premier of China Zhou Jiahua wrote an inscription for the State Information Center: "Make a good job of economic development prediction so that it can serve a socialist market economic system". It also published all kinds of economic prediction vehicles such as *World Economic Prospect, Economic Prediction Report, Chinese Market Prospect,* and *Chinese Automobile Market Prospect.* The central government of China attached great importance to these prediction reports and wrote many comments on them. For instance, Vice-Premier Zhou Jiahua once commented on one issue of *Economic Prediction Report* as: "Very good. Send every issue of this report to the office of General Sectary and Premier Li".

Second, the Institute of Quantitative Economics and Technical Economics at the Chinese Academy of Social Sciences began to release the *Economic Bluebook*, namely, the *Chinese Economic Situation Analysis and Prediction* on an annual basis since 1991. It was compiled at Economic Situation Analysis Conference based on the articles and views of conventioneers from related government departments, institutions of higher learning and research institutes.

Finally, the Institute of Macroeconomics at the State Development and Reform Commission also made annual economic situation analysis and prediction reports, either to be submitted to the Central Party Committee and State Council or for public publication.

China's economic prediction is making constant progress. As far as the quality of prediction on economic growth rate is concerned, the accuracy in economic prediction of China's major prediction institutions has improved. For example, China's real economic growth rate in 2006 was 10.7%. However, there was a great divergence in the prediction made by over 40 domestic and foreign institutions and scholars in 2005. In the case of an average error rate of 12.5%, the predicted value provided by Prediction Department of State Information Center had a relative error rate of 0.93%. Given that a prediction with an error rate lower than 5% is an accurate prediction. Also, in 2007, China's real economic growth rate was 11.4%. Predictions made by 42 domestic and foreign institutions and scholars in 2006 were strikingly different. In the case of an average error rate of 14.3%, the predicted values provided by Prediction Department of State Information Center and Institute of Quantitative Economics and

Technical Economics were 10.9% with a relative error rate of 4.3%. Since the relative error rate was lower than 5%, their prediction was considered as an accurate one. In fact, it was the most accurate one among the six accurate predictions (see Wang Liyong's "An Evaluation and Analysis of the accuracy of Prediction on China's Economic Growth in 2007" in the second issue of *Economics Dynamics in 2008*).

To improve the quality of economic prediction, we need to reinforce the training of economic prediction workers. One of the best ways of training is to summarize and advocate experience and feelings of economic prediction professionals while introducing quantitative economic theory and methods. In China, there are more than one thousand economic prediction professionals. In order to systemize their predictions, State Information Center arranged for them to compile and publish a book, *Economic Prediction and Case Analysis in the 1990s* (edited by Wu Weiyang and published by China Economic Press in 1995). The preface of this book was written by Wu Jiapei and the book discussed the following 6 issues: (1) Why is economic prediction possible? (2) What is the method of economic prediction? (3) What are the tools of economic prediction? (4) Are there any good practices in economic prediction? (5) What is the major function of economic prediction to economic planning, decision-making, control and management? (6) Contemporary economic prediction science includes new advances in quantitative economics. What is most useful is that this book also contains 22 cases on market prediction, macroeconomic prediction and strategic development prediction and an analysis of them. The book played an important role in economic prediction training at that time and it is still a valuable economic prediction work.

Economic prediction theory and practice advance with the times. This is because the law of economic activity is not static. In fact, volatility, relevance and tendency of economic development vary from time to time and from place to place. To meet the new demands of economic development, we need to make more and greater efforts in economic prediction. In the late 20th century and the early 21st century, the State Information Center made constant efforts in reinforcing the foundation work of economic prediction including the construction of economic database and design of economic mathematical models.

More than 10 key databases and networks were constructed, such as Chinese Economic Network Statistics Database, Real Estate Information Database, Retail Market Monitoring Database of Consumer Electronics and Chinese Economic Private Network. A great number of large-scale models were built such as Chinese Macro-Economic Model, China–Japan Trade Integration Model, Multi-Sector Price Model, Regional Economic Development Integration Model, Economic Prosperity Analysis and Consumer Intention Analytical Model and Integrated Quantitative Analysis Simulation Model that combines population, economy and environment. This model can be used to study the measurement of environmental pollution and its treatment. Apart from these two aspects, various platforms that integrate data, model and application were developed, like macroeconomic monitoring, prediction and analytical platform introduced in 2006 and the China regional monitoring platform introduced in 2007. These platforms not only created a favorable condition for the State Information Center to conduct economic prediction, but also facilitated the prediction of the Chinese economy for domestic and foreign institutions and scholars.

3. Overcome Difficulties and Promote Development

To conduct economic prediction is to learn from the past and predict the future. Since the past has become a fact whereas the future is full of uncertainties, it is easier to learn from past than predict the future. There are two kinds of difficulties hindering our economic prediction:

The first kind of difficulties is general difficulties that all predictions are confronted with:

(1) Economic prediction is different from the weather forecast in that the economic development process is dependent upon human will and actions, and involves various contradictions and conflicts in interests. It is very hard for predictors to monitor all incidents occurring in an economic process and all operational factors as well as maintain their objectiveness without any subjectivity when making assumptions about what will happen or what will not happen on the other hand.

(2) Economic prediction is both a science and an art. To make sure that economic prediction is repeatable and can withstand scrutiny,

economic predictors must receive professional training and possess expertise. It is by no means an easy task to create a quality economic prediction team.

The second kind of difficulty is special difficulties that are related to the unique characteristics of China's economy and Chinese economic prediction:

(1) The Chinese economy is different from that in developed European countries and the United States. China is a major developing country and its economy is in transition. During this process, the influence of systematic and policy factors are significant. Its economic development process is filled with inflexion points and emergencies; although we strive for a steady development-fluctuations are by no means insignificant. As a matter of fact, marked ups and downs occur frequently, which will definitely make it more difficult to make economic predictions.

(2) Because of China's deepening reforms and opening-up, the acceleration of integration into the global economy, greater economic dependence on finance and uncertainty and unprecedentedly huge uncertainty and instability in economic development under the impact of global financial crisis, it is more difficult than ever to make economic prediction.

(3) China's economic prediction does not have a long history or a strong basis. Our input in economic prediction as well as support from economic data, and prediction theory and methods were insufficient. Conducting economic prediction is costly and not always accurate as it is hard to make quality economic predictions all the time. In spite of this, people are too satisfied with general analyses of the economic situation and intuitive predictions on economic development trends to further study different kinds of prediction theories and familiarize themselves with various prediction methods.

(4) It is difficult to stabilize and expand the economic prediction team. On the one hand, well-informed prediction workers are likely to be promoted and leave their prediction team if they make good economic predictions and attain certain achievements; on the other hand, ordinary prediction workers are likely to leave their prediction team due to poor work conditions.

However, no difficulty is insurmountable. In the long march of China's economic prediction industry, we should try our best to overcome all difficulties and strive for a final victory. To promote a sound and rapid development of our economic prediction, we need to make efforts in these aspects:

First, work at economic prediction as an undertaking and a profession and concentrate our strength on building a quality national economic prediction team and an economic prediction network that covers all professions and regions. Although with widespread economic models, computers and the advanced functions of software, every department and individual can make economic prediction and decide his/her own action. The entire country cannot afford to have a reliable and well-developed economic prediction center as its basis for economic decision-making and control against the background of globalization and fast growing informatization. Therefore, it is an urgent task to consolidate the national economic prediction team.

Second, develop economic prediction theory and methods. As the saying goes, a handy tool makes a handy man. We cannot be content with traditional economic prediction theories and methods such as the input–output model and econometric model. Instead, we should study and employ modern economic prediction theories and methods such as the neural network model that caters for the needs of nonlinear economic development prediction, wavelet analysis theory and wavelet network that combines neural network model and wavelet analysis theory.

Third, develop an economic prediction consultation industry as an important modern service industry. Economic prediction should serve both the government and the public. On the one hand, the development of the economic prediction consultation industry mainly relies on national economic prediction organization and its network. On the other hand, it relies on the commercialization and industrialization of economic prediction and its results. In return, the development of the economic prediction consultation industry will promote economic prediction that serves the government.

A Review of 10 Years of Economic Forecasting in China[*]

Wang Tongsan[†] and Shen Lisheng

1. The Origin of Annual Forecast Model of Chinese Macro-Economy

Quantitative economics has played an increasingly important role in the transformation from a traditional planned economy to a market economy since the reform and opening-up in China in the 1980s. In its field of study, to forecast and analyze the macro-economy by adopting an economic model is one of the main objectives.

In 1987, a cooperative research project called the China–US–Japan Link Model was co-established by the Institute of Quantitative and Technical Economics at the Chinese Academy of Social Sciences, and Professor Lawrence Klein at the University of Pennsylvania and Professor Liu Zunyi at Stanford University. The project was sponsored by the Ford Foundation. With the guidance and kind help of the two professors, we were able to establish an annual forecast model of the Chinese macro-economy. Apart from applying it to the Link Model, we can also use it to simulate, analyze and forecast the economic situation in China. The model was first officially applied to the macro-economic forecast and a forecast report was then released in the autumn of 1990, after an internal trial run was launched upon its establishment in the same year. It has been the formal forecast tool for economic forecast for 10 years since the foundation of the Analysis and Forecast of Economic Situation Project at the Chinese Academy of

[*]This article was originally contained in *The Frontier of Quantitative Economics*, published by Social Sciences Academic Press in 2001.
[†]Wang Tongsan, the member of the Chinese Academy of Social Sciences and Director, Researcher and Doctoral Supervisor of Institute of Quantitative and Technical Economics.

Social Sciences in 1991. On the 20th anniversary of the Training Class at the Summer Palace[1], let us review the development of economic forecast in the past 10 years.

2. The Evolution of the Economic Forecast Model

The Chinese macro-economic model, founded in the mid-1980s, is basically a supply-oriented model. It contains 8 modules, 220 equations, 220 endogenous variables and 35 exogenous variables.

The Chinese Economic Model has been updated annually since it was officially put into application. New statistics of the past year are added and equations in modules re-estimated during every update. At times, some equations are modified according to the change in situation. Significant changes are mainly 2-fold.

2.1. *Changes in the aspect of statistical system*

Changes in the statistical indicator system depend upon those in the system of national accounts. As is known to us all, borrowed from the Soviet Union in the 1950s, China's system of national accounts is in fact the System of Material Product Balances (MPS), a product that corresponds to the planned economic system. The MPS has numerous defects. For instance, it employs the total product of society, i.e. the aggregation of the total value output of various material production departments, to calculate the sheer size of national economy. However, repetitive computation occurs for the total value of output which includes the transformation value of raw material. To avoid this, the net output value of each material production department and the sum of the values, or in other words the national income, are brought into the statistical indicators as well. Another disadvantage of MPS is the lack of statistics about non-material production departments and a complete statistics on the service sector.

---

[1]In the summer of 1980, the American economic delegation headed by Professor Lawrence Kline, held a 7-week-long Econometrics Training Class in the Summer Palace. It was a crash course for the attendees but, however, an event of significance for Chinese economic study, which was left far behind by modern economics in the rest of the world. The Chinese Association of Quantitative Economics has celebrated the 10th and 20th anniversaries of that event.

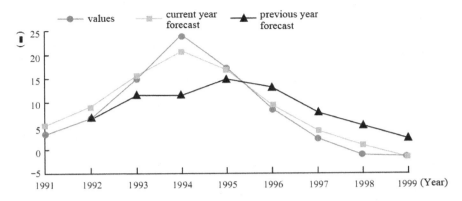

Graph 3. Statistical and Predicated Values of Consumer Price Index

rate, whether it is the trend of the current year forecast, of the previous year forecast, and of the differences between current year forecast and previous year forecast.

Graph 3 is a comparison between statistical values and two predicated values of consumer price index. It shows that (1) the projection error of the current year forecast is small while that of the previous forecast is big, and (2) the current predicated values of all those years except 1994 are higher than actual statistics, and at the meantime the previous predicated values are not higher than statistics until 1996, being consistent with the prediction about GDP growth rate.

Graph 4 makes a comparison between statistical values of total social retail sales and their predicated values. Generally speaking, all the other predicated values are in close proximity to the actual statistics other than the previous year predicated values in 1998, for which the high prediction about GDP growth rate of that year accounts.

Graph 5 compares two predicated values of total import with its statistical values. Foreign trade conditions have not been predicted until the autumn of 1992, thus comparison begins at the year of 1993 as to the total import.

Graph 6 concerns the comparison between statistical and two predicated values with regard to the total exports from 1993 for the same reason as the total imports. From both the forecast of imports and exports, we can see that the current year forecast is closer to statistics than the previous year

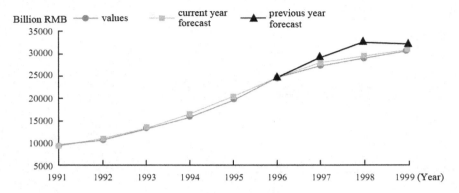

Graph 4. Statistical and Predicated Values of Total Social Retail Sales

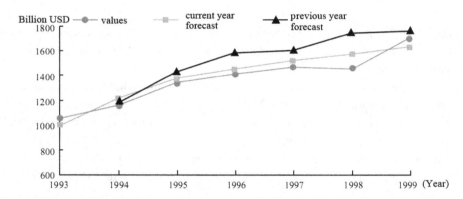

Graph 5. Statistical and Predicated Values of Total Import

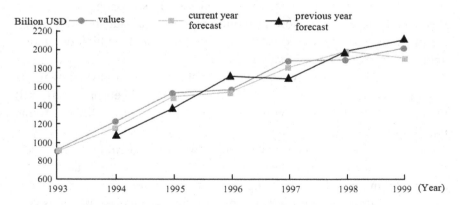

Graph 6. Statistical and Predicated Values of Total Export

forecast. But the difference is that the predicated values of total imports are practically higher than the statistical ones, while no regular pattern is detected as to the projection error of the total export forecast. This is because an overestimation of imports comes along with that of domestic economic development as the economic situation of a nation determines its imports. On the other hand, given that exports change with the fluctuations of the international economy, the irregular pattern of export forecast error shows that our judgments on the global economic situation are not tendentious.

In summary, the current year forecast is more accurate than the previous year forecast in situations of all 6 indicators. Two reasons may account for this. Firstly, the economic situations of the first half year and government policies implemented in the past half year are already clear at the time when the current year forecast is to be made. In this case, the error of projection will not be big. Secondly, actual statistics of major economic indicators have been published in *A Statistical Survey of China* every autumn before the current year forecast. Helpful information from them will be added to the model, which contributes to a small prediction error.

What is more, it is interesting to find that predicated values were generally below the level of actual statistics until 1995, while it turns to be the other way around thereafter. This is actually proof of the influence of the inertia of trend extrapolation when using an economic model to forecast. The influences are 2-fold:

The first influence is from the trend extrapolation inertia that existed in the model. The economic model, the econometric model in particular, predicts the future to some extent based on information from the past and the present. In consequence, unprocessed model predictions will in a way demonstrate the inertia of trend extrapolation. For example, even though China in its early 1990s suffered economic retrenchment, the economic growth rate being relatively low, unexpectedly the economy became overheated from 1993 which, however, failed to be predicted via the model from 1993 to 1995. Another example is that affected by the trend extrapolation inertia of high macro-economic growth rate during 1993 to 1995, predication values after 1995 tended to be higher than actual situation where China has remained in a down economy due to the Asian financial crisis since 1996's "soft landing".

The second influential factor is the inertial thinking of predictors. Economic forecasting, though based on economic models, is a process of "science plus art", as predicators have to amend the calculation results by taking account of potential influences by elements that are neglected in the model. During forecasting, this processing pattern of "model calculation — analytical amendment" may function more than once. And in general, predicators' inertial thinking often results in a smaller amount of variation in predications than that of the real situation.

For instance, the year before 1995, China had experienced a consecutive high two-digit growth. However our subjective judgment always seemed to lag behind reality, thinking that the economic growth and inflation had already reached the top and expecting that the economy should slow down next year. Predicted values appeared lower than real statistics due to the misjudgment. Then facing the continuous downturn since 1996, we thought that economy would embrace a rising period in the following year, which led to high predications. The influence on predication results by subjective judgments is obvious and essential. Therefore, it requires a good control and usage of subjectivity during the forecasts of researchers.

In spite of all this, we are happy to find that the model has kept consistent even though having been revised several times. The changes of the main economic indicators, as we can see from the comparative graphs above, show consistency. When the predicted values of the GDP growth rate vary, identical changes occur to that of fixed investments and of the consumer price index. This verifies the good functioning of the economic simulation operation of the model and also makes it possible for the model to be employed as a tool of policy analysis, and if the policy analysis is deemed reliable, the performance of an economic model can be said to be excellent.

4. Reflections on the Economic Model and Forecast

We have harvested precious experiences from 10 years of forecasting, which we summarize as follows:

(a) Economic forecasting is on the basis of the economic model and priority should be given to improvement of the model.

The current situation is that China is in a transitional period toward a market economy and thus the operative economic system is changing, which calls for a continuous amendment of our own model. That includes transforming the short-term prediction oriented model into a demand oriented one and applying new research results of econometric methods to the amelioration of the model's function and quality.

(b) Economic forecasting, especially short-term forecasting, should be improved so as to be capable of judging correctly the macro-economic situation.

The following steps should be duly done: (1) Keep a close eye on the development of the macroeconomy; (2) communicate more with other predicators, officials relevant to economy, and enterprises; (3) supplement and verify the predictions by using methods other than the economic model.

(c) The forecasting about the turning point of economic development deserves academic attention.

Contrary to the experience of developed market economy nations, fluctuations of the economy in China are large. Despite that changes over two adjacent years have narrowed since 1992, the gap between a high increase year and a low increase year is considerable. For instance, the GDP growth rate of 1992 was twice as that of 1999, the gap in between being seven percentage points. This is because China is heading toward a mature market economic system and there is still a long way to go. As a result, economic forecasting is definitely not an easy task and the study of the economic turning point in particular is challenging but important.

(d) We should learn to better understand and estimate the international economic situation.

China's economy has been increasingly closely related to the global economy ever since the policy of reform and opening-up. As a result, the impact of economic globalization on China' economy has only become greater. The influence, though indirect, of the East Asian financial crisis in 1997 on China cannot be ignored either. Nations in East Asia were beset by currency devaluation, negative growth of the economy, and decreased international trade, due to which China's imports and exports to Asia dropped significantly. This is understandable as China's trade with Asian countries takes up half of its total foreign trade. However, China began to benefit from the ending of the crisis in 1999 and the rejuvenation of the region's economy. In the first half of 2000, the total imports and exports in China increased by 37.3%, offering a boost for China to get rid of seven years of economic downturn.

Obviously, the economic relations between China and the world will be more and more close, which means China's economy is to be more vulnerable to the global economic situation. Thus it is necessary to pay more attention to international trade in our economic model and the impact of foreign investment and trade on China's economy.

(e) As the economic system is evolving in China, how to use statistics in the planned economy for reference when forecasting in the market economy is a research issue worth studying.

Take the labor market as an example. Under the circumstances of a planned economic system, the labor force in urban areas is distributed by the department in charge of state-owned enterprises and is to be accepted unconditionally. Consequently, the number of workers for a particular task may surpass that actually needed. In other words, labor productivity is artificially reduced and surplus labor, or hidden unemployment, in state-owned enterprises is common place. However, a market economy requires enterprises to hire workers according to their needs. In order to participate in market competition, surplus labor has to be laid off so that working efficiency and labor productivity in the enterprise can be raised, which is unavoidable during enterprise reform. Hence to make use of the employment data in a planned economy, it is essential to properly estimate the hidden unemployment and exclude it.

One-Way Effect Causal Relationship of GDP and Energy Consumption in Taiwan[*]

Chen Yanwu, Yao Feng and Wu Chengye[†]

1. Literature Review

Granger (1969) introduced a now widely applied conception of non-causality in order to decide the direction of causality in a time series. Sims (1972) discussed causality statistical tests during the moving average process via looking at the causality between currency and income in America. Significant achievements have been made in the empirical study of macro-economic, fiscal and financial domains in the past decades, although most of them are still limited to Granger's non-causality. Hosoya (1991) effectively described the relations between two sets of sequential time variations in a second order stationary process with a non-deterministic trend, which laid a theoretical foundation for one-way causality analysis. Yao & Hosoya (2000) put forward the Wald test on one-way causality measure theory, based on co-integration process ECM time sequence's causality, which fundamentally solved the problem of applying one-way causality measure theory to statistical test of dynamic economic system analysis. Yao & Shi (2003) were the first to apply this method to empirical analysis of Japan–China economic relations. This chapter attempts to

*Foundation Supported Projects: Specialized Research Foundation for Doctor Stations in Universities (20050385001), 2007 Fujian Provincial Support Project for Talents in Colleges and Universities, and MEXT Core Research Foundation B (16402023).

†Chen Yanwu, Doctor, Associate Professor, College of Commerce, Academy of Quantitative Economics, at Huaqiao University; Yao Feng, Doctor, Professor, Branch of Economics, Kagawa University; Wu Chengye, Doctoral Supervisor, Professor, Academy of Quantitative Economics, Huaqiao University.

reveal the causal relationship between economic development and energy consumption in Taiwan through Wald statistics of one-way causality measure.

There has not been a consensus on causal relationship between energy consumption and economic growth in various nations during different periods since the end of 1970s. The innovative research of Kraft & Kraft (1978) on the causal relationship between American GDP and energy consumption showed that from 1947 to 1974 economic growth in America influenced energy consumption but not the other way around. This result tells us that energy protection will not have a negative effect on economic development. Yu & Jin (2002) analyzed the co-integration relation between every two years of energy consumption, national income and employment based on monthly statistics from January 1974 to April 1990. They found that co-integration relation did not existed in any pair of the three elements, or in other words, that energy consumption had kept a mutual relationship with both national income and employment and thus will not affect economic growth. Uri (1995) argued that the lack of resources had an influence on economic growth of America, and Uri (1996) pointed out that price fluctuations of crude oil will affect employment and unemployment rate. His opinions encouraged a large number of studies on the relations between energy and economic growth.

As for research on relations between energy consumption and economic growth in Taiwan, Cheng & Lai (1997) were the first to adopt Granger's causality method mentioned by Hsiao (1981) to study the relationship between GDP and energy consumption in Taiwan from 1955 to 1993. They found that GDP influenced energy consumption while the latter did not affect the former and pointed out that increasing economic growth required a large amount of energy. Based on statistics about GDP and energy consumption in Taiwan from 1954 to 1997, Yang (2000) studied the causal relationship between GDP and energy consumption as well as its constituents like coal, oil, natural gas and electricity. The study demonstrated the interplay of energy consumption and GDP and the restriction of energy shortage on economic growth. In addition, GDP and coal consumption influenced each other and so did GDP and electricity. GDP influenced oil but not the other way around, while natural gas affected GDP unidirectionally. However, Lee & Chang (2005) got different results in their research on causal relationship between GDP and energy consumption and its constituents in Taiwan from 1954 to 2003. Their study concluded that energy could be a long-term driving force for the economy but energy protection would hinder economic growth. All the literature mentioned above failed to reach an agreement on the direction of causality relations, although all studied the same variables at different time periods.

The 1990s witnessed an increase of intermediate speed and comparative steadiness in the economy of Taiwan. However, in 2001 Taiwan suffered a sudden economic recession due to the global economic situation, and exacerbated by problems in its interior political structure. Since then 10 years of intermediate-speed economic growth has encountered great challenges, leading Taiwan's economy to a turning point. As for the relationship between economic growth and energy consumption, the average growth rate of economy from 1980 to 1996 reached 7.59% and that of energy consumption was 5.84%, showing a relatively low energy consumption level in economic output. From 1997 to 2002, economic growth decreased to 3.63% while energy consumption rate was still 5.58%, which demonstrated a declining energy rate in Taiwan. Excessive energy consumption could not effectively boost the economy and would result in an imbalance between energy consumption and economic growth. In recent years, the international crude oil price has been rocketing, which increases the energy risk for Taiwan as 97% of energy needed is imported. In this case, it is of realistic significance to probe into the relationship between economic growth and energy consumption and its constituents. This chapter analyzes annual statistics of GDP and energy consumption in Taiwan from 1954 to 2005 by adopting the one-way causality measure theory so that the co-integration relationship and one-way causality measure between GDP and energy consumption with its constituents such as oil and electricity can be revealed.

2. One-Way Effect Causality Measure Theory

To begin with, the analytical method of one-way effect causality measure for a time sequence produced by an error correction model (1) is introduced. Then we analyzed the causal relationships based on various measure theories by making use of Wald statistics of one-way effect causality measures. The calculation process is presented and a confidential interval of measures is set up via statistics tests.

A linear finite order p-dimensional VAR (a) process can be expressed in an error correction model:

$$\Delta Z(t) = \alpha \beta' Z(t-1) + \sum_{k=1}^{a-1} \Gamma(k) \Delta Z(t-k) + \mu + \varepsilon(t) \qquad (1)$$

In Equation (1), $\varepsilon(t) \sim i.i.d.$, $N(0, \Sigma)$ $(t = 1, 2, \ldots, T)$, and μ is p-dimensional vector. α, β are $p \times r$ order non-singular matrix $(r \le p)$. α is called adjustment coefficient matrix, representing the adjustment speed of time sequence from non-equilibrium to equilibrium state. β is named as co-integration parameter matrix, showing the long-term balance relationship among time sequences. The primary observed values of $Z(t)$ are original values. Then we can determine the co-integration numbers and vectors through applying complete information maximum likelihood ratio test brought up by Johansen (1991).

Based on an error correction model, FMO[1] and OMO[2] of non-stationary time series are determined and tested as follows: As to model (1) and time sequence $Z(t) = (X(t)'Y(t)')'$, of which $X(t)$ and $Y(t)$ are p_1 and p_2 vectors $(p = p_1 + p_2)$. After setting up co-integration order r and co-integration vector $\hat{\beta}$, we define in model (1) that $\Pi = \alpha \beta'$ and meanwhile estimate by ordinary least squares all other parameters, including $\hat{\alpha}$, $\hat{\Gamma}(k)$ $(k = 1, 2, \ldots, a)$ and covariance matrix Σ of model error term.

Define Σ_{ij} $(i, j = 1, 2)$ as p_1 and p_2 order partitioned matrices of Σ, $C(e^{-i\lambda})$ as adjoint matrix of plural polynomial matrix

$$I_p - (I_p + \alpha \beta') e^{-i\lambda} - \sum_{j=1}^{4} \Gamma(j) \left(e^{-ji\lambda} - e^{-i(j+1)\lambda} \right)$$

and $\Lambda(e^{-i\lambda}) = C(e^{-i\lambda}) \Sigma^{1/2}$ is frequency spectrum function.

$$f(\lambda) = \frac{1}{2\pi} \Lambda(e^{-i\lambda}) \Lambda(e^{-i\lambda})' = \begin{bmatrix} f_{11}(\lambda) & f_{12}(\lambda) \\ f_{21}(\lambda) & f_{22}(\lambda) \end{bmatrix} \qquad (2)$$

---

[1]Frequency-wise measure of one-way effect.
[2]Overall measure of one-way effect.

Based on Equation (2), the one-way frequency spectrum measure of time sequence $\{Y(t)\}$ to $\{X(t)\}$ can be defined as:

$$M_{Y \to X}(\lambda) = \log\left(\det f_{11}(\lambda)/\det\left\{f_{11}(\lambda) - \tilde{f}_{12}\tilde{f}_{22}^{-1}\tilde{f}_{21}\right\}\right) \qquad (3)$$

In Equation (3),

$$\tilde{f}_{11}(\lambda) = f_{(11)}(\lambda), \quad \tilde{f}_{21}(\lambda) = \left(-\sum_{21}\sum_{11}^{-1}, I_{p2}\right)C(e^{-i\lambda})^{-1}f_{.1}(\lambda)$$

$f_{.1}(\lambda)$ is the original p_1 column of matrix $f(\lambda)$, which is

$$\tilde{f}_{22}(\lambda) = \frac{1}{2\pi}\left(\sum_{22} - \sum_{21}\sum_{11}^{-1}\sum_{12}\right)$$

Rearrange the parameters in model (1) into

$$\theta = vec\beta', \quad \psi = vec\left(vec(\alpha, \Gamma), v\left(\sum\right)\right)$$

$v(\Sigma)$ is the vector of $n_\psi = p(r + p(a - 1)) + p(p + 1)/2$, $\Gamma = \{\Gamma(1), \ldots, \Gamma(a)\}$, therefore, the one-way OMO of $\{Y(t)\}$ to $\{X(t)\}$ can be defined as:

$$G(\theta, \psi) = \frac{1}{\pi}\int_0^\pi M_{Y \to X}(\lambda|\theta, \psi)d\lambda \qquad (4)$$

In this case, we assume $G(\theta, \psi)$ is the differentiable function of parameter (θ, ψ).

Then we constructed the Wald statistics of one-way effect causality measure $G(\theta, \psi)$:

$$W = T\{G(\hat{\theta}, \hat{\psi}) - G(\theta, \psi)\}^2/H(\hat{\theta}, \hat{\psi}) \qquad (5)$$

χ^2 distribution with 1 Asymptotic degree of freedom, of which $(\hat{\theta}, \hat{\psi})$ is the maximum likelihood estimation of model parameter (θ, ψ), and $H(\hat{\theta}, \hat{\psi})$ the variance–covariance matrix of $\sqrt{T}\{G(\hat{\theta}, \hat{\psi}) - G(\theta, \psi)\}$.

The confidential interval of $(1 - \alpha\%)$ of causality measure $G(\hat{\theta}, \hat{\psi})$ is:

$$\left[G(\hat{\theta}, \hat{\psi}) - \sqrt{\frac{1}{T}H(\hat{\theta}, \hat{\psi})\chi_\alpha^2(1)}, \; G(\hat{\theta}, \hat{\psi}) + \sqrt{\frac{1}{T}H(\hat{\theta}, \hat{\psi})\chi_\alpha^2(1)}\right] \qquad (6)$$

Thus the annual statistics of 1952 that we discussed before can hardly be counted as a large-sample issue. During the calculation we replaced numbers of sample T with $T - n_\psi$.

3. Preliminary Analysis

3.1. *Statistics and unit root test*

First we carry out a unite root test on the annual statistics for GDP, energy consumption and its constituents like oil and electricity consumption in Taiwan and then construct an error correction model applying to one-way effect causality analysis. We used the annual statistics from Taiwan from 1954 to 2005. Variables adopted include: GDP of Taiwan (one million New Taiwan Dollars, 2001 as the base year), energy consumption (ENE), oil consumption (OIL), and electricity consumption (ELE). Oil is set up as the labeled amount in the energy consumption series, the unit being kiloliter. Model analysis adopts usually natural logarithm values of each indicator. In considering that coal consumption and that of its relevant products and gas consumption have taken up a low proportion in overall energy consumption, the paper focuses on the analysis of the co-integration relation and causal relationship between GDP and ENE, OIL and ELE.

Table 1 shows all series unit rook test results, including that of GDP, ENE, OIL and ELE. From Table 1, we can see that GDP, ENE, OIL and ELE belong to one order integration I(1). Take GDP time series as example, t value of time trend term is -0.3048, greater than ADF statistics

Table 1. ADF Unit Root Test

Variables	T values	DW values	Conclusion
GDP	−0.3048	1.9026	non-stationary
ENE	−0.6783	1.7874	non-stationary
OIL	1.3373	1.8183	non-stationary
ELE	2.0811	1.7529	non-stationary
ΔGDP	−6.5946	2.0108	stationary
ΔENE	−6.0809	1.9848	stationary
ΔOIL	−6.2508	1.9801	stationary
ΔELE	−5.5868	2.0247	stationary

−3.5005 and statistically significant. It shows that the GDP time series is a deterministic trend non-stationary process. After further differential treatment, the t value of time trend is −6.5946, less than ADF statistics of −3.5024 and not statistically significant. It demonstrates that ∆GDP is a stationary process. The ENE, OIL and ELE series unit root tests are carried out in the same way. The results show that ENE, OIL and ELE time-series are all deterministic trend non-stationary processes and they change into stationary processes after differential treatment.

3.2. *Order selection and non-autocorrelation test*

To establish the vector autoregression model, not only the stationary condition needs satisfying but also the lag phase should be determined. As for the selection of lag orders, we comply with the Akaike Information Criterion (AIC):

$$\text{AIC} = \log\left(\frac{\sum_{t=1}^{N} u_t^2}{N}\right) + \frac{2k}{N}$$

In this equation, U_t represents residual error, N sample size and k maximum lag phase. The principle of selecting k is to minimize the AIC value during increasing k value. According to AIC, the best model for the two pairs of variables GDP and ENE, and GDP and ELE is the error correction model (ECM) (1) of lag 1 order. The best model for variables GDP and OIL is the ECM (3) of lag 3 order. The estimated AIC statistics are summarized in Table 2.

To select the proper model, both the co-integration order r should be determined based on $\Gamma(r)$ statistics, and a non-autocorrelation test of residual series need further carrying out. As for the non-autocorrelation

Table 2. AIC Values

Variables	AIC (1)	AIC (2)	AIC (3)	AIC (4)
GDP and ENE	−8.2786	−8.2529	−8.0805	−7.9752
GDP and OIL	−6.7237	−6.7414	−7.0864	−7.0242
GDP and ELE	−8.6167	−8.5506	−8.4042	−8.4954

test of residual series, we adopt Hosking (1980) statistics which is more effective to small sample issues:

$$\text{Hg}(s) = T^2 \sum_{j=1}^{8} \frac{1}{T-j} \, tr \left\{ \hat{C}_{0j} \hat{C}_{00}^{-1} \hat{C}_{0j}' \hat{C}_{00}^{-1} \right\}$$

In the equation, $\hat{C}_{0j} = T^{-1} \sum_{t=j+1}^{T} \hat{\varepsilon}_t \hat{\varepsilon}_{t-j}'$. Under the assumption of the non-autocorrelation test of residual series, as to sufficiently large sample number T and sufficiently large order number $s(s \gg a)$, the asymptotic degree of freedom of autocorrelation test statistics $\text{Hg}(s)$ is the x^2 distribution of $f = p^2(s-a)$ and a is the lag order number of the model. In the analysis of the two variable models, we determine $s = 18$ when it comes to the sample number and order number of the vector autoregression model. Under the assumption of non-autocorrelation test of residual series, the observed value of test statistics $\text{Hg}(s)$ and p-value are summarized in Table 3. The statistical test shows that all residual terms of the model that is recognized in the selected lag length will satisfy the non-autocorrelation at 95% of the confidence level. This allows us the implementation of one-way effect causal analysis on the basis of proposed ECM.

3.3. *Co-integration relationship*

Based on the ECM of two variables' autoregression model, Table 4 listed out the characteristic roots, characteristic vectors and estimated values of trace statistics. The critical value of trace statistics τ of the co-integration order test is given by MacKinnon *et al.* (1999). In this chapter, we estimate co-integration order numbers r according to not only $\hat{\tau}(r)$, but also statistics, relevant models and other aspects. See more in Johansen (1996), Yao & Hosoya (2000).

Table 3. Hg(s) Values in Non-Autocorrelation Test of Residual Series

Variables	Hg-statistic	p-value
GDP and ENE	70.6165	0.266
GDP and OIL	64.7613	0.198
GDP and ELE	82.7351	0.058

Note: Static value Hg(s) is defined by Model (13).

Granger, C.W.J. (1969). Investigating causal relations by cross-spectrum methods, *Econometrica*, 39(3), 424–438.

Hosking, J.R.M. (1980). The multivariate Portmanteau statistics, *Journal of the American Statistical Association*, 75, 602–608.

Hosoya, Y. (1991). The decomposition and measurement of the interdependency between secondorder stationary processes, *Probability Theory and Related Fields*, 88, 429–444.

Hsiao, C. (1981). Autoregressive modeling and money income causality detection, *Journal of Monetary Economics*, 7, 85–106.

Johansen, S. (1996). *Likelihood-Based Inference in Cointegrated Autoregressive Models*. Oxford University Press.

Kraft, J. & Kraft, A. (1978). On the relationship between energy and GNP, *Journal of Energy and Development*, 3, 401–403.

Lee, C.C. & Chang, C.P. (2005). Structural breaks, energy consumption, and economic growth revisited: Evidence from Taiwan, *Energy Economics*, 27, 857–872.

Mackinnon, J.G., Haug, A.A. & Michelis, L. (1999). Numerical distribution functions of likelihood ratio tests for cointegration, *Journal of Applied Econometrics*, 14, 563–577.

Osterwald-Lenum, M. (1992). A note with quantiles of the asymptotic distribution of the maximum likelihood cointegration rank test statistics, *Oxford Bulletin of Economics and Statistics*, 54(3), 461–472.

Sims, C.A. (1972). Money, income and causality, *American Economic Review*, 62, 540–552.

Uri, N.D. (1995). A reconsideration of the effect of energy scarcity on economic growth, *Energy*, 20, 1–12.

Uri, N.D. (1996). Crude oil price volatility and unemployment in the United States, *Energy*, 21, 29–38.

Yang, H.Y. (2000). A note on the causal relationship between energy and GDP in Taiwan, *Energy Economics*, 22(3), 309–317.

Yao, F. & Hosoya, Y. (2000). Inference on one-way effect and evidence in Japanese macroeconomic data, *Journal of Econometrics*, 98(2), 225–255.

Yu, E.S.H. & Jin, J.C. (2002). Cointegration tests of energy consumption, income and employment, Resources and Energy, 14, 259–266.

姚峰，史宁中，《日本经济发展与中日贸易的计量分析》，《管理科学学报》，2003年第 4 期 (Yao, F. & Shi Ningzhong (2003). "Japanese Economic Growth and Quantitative Analysis of Trade between China–Japan", *Journal of Management Sciences in China*, 4.).

Part Two

Record of Major Events in the 30-Year History of the Chinese Association of Quantitative Economics (1979–2009)*

Peng Zhan et al.

1979

On March 30, 1979, the China Quantitative Economics Society[1] was established in Beijing.

From March 22–30, 1979, 18 scholars and experts from across China held a discussion in Beijing on the main issues concerning quantitative economics, including (1) the necessity to lay emphasis on quantity and method in economic study; (2) approaches to evaluate bourgeois econometrics; (3) the naming of the discipline; (4) ways of using quantitative economics to serve the four modernizations[2]; (5) mathematic approaches that can be applied to economic work; and (6) the founding of an association of quantitative economics and working assumptions.

---

*Due to limited time and resources, this record of major events is incomplete, especially on specialized committees and local associations, nor specific enough or totally exact. Information and materials are sincerely welcomed and expected so that the record can be revised and supplemented. Feel free to contact the secretariat of Chinese Association of Quantitative Economics if there is any.

[1] Now the Chinese Association of Quantitative Economics.

[2] Four modernizations refer to the modernization of agriculture, industry, national defense and science and technology. This conception was first put forward as a government policy to construct the new China by former Premier Zhou Enlai at the third National People's Congress in December, 1964.

1980

From June 23 to August 9, 1980, commissioned by the Chinese Academy of Social Sciences, the China Quantitative Economics Society held the Econometrics Workshop in Summer Palace, Beijing. Seven American scholars headed Professor Klein lectured and 100 trainees attended.

From August 15–16, 1980, the Quantitative Economics Symposium was convened in the Summer Palace, Beijing. Scholars and journalists discussed ways to conduct the research, application and teaching of quantitative economics.

From 1980 to 1981, thanks to the help of experts from the Institute of Economics at the Chinese Academy of Social Sciences and School of Planning and Statistics[3] at Renmin University of China, the first regional input–output table in China was made in Shanxi Province, which included a physical table of 88 kinds of products in 1979 and a value table of 27 departments in the same year.

In 1980, lectures of the 7 professors namely Klein, Anderson, Ando, Gregory Chow, Lawrence Liu, Vincent Su and Cheng Hsiao at the Econometrics Workshop in the Summer Palace were compiled into *Econometrics Handouts* by Wang Hongchang and the others and later published by Beihang University Press.

1982

From February 28 to March 3, 1982, the Quantitative Economics Symposium of China (and the First Annual Meeting) was convened in Xi'an by the China Quantitative Economics Society jointly with relevant organizations, with more than 160 attendees including representatives and staff, and over one hundred dissertations were presented. At the meeting, Mr. Xu Dixin delivered the opening statement, followed by speeches by Mr. Ma Hong and Mr. Yu Guangyuan respectively, and a pre-recorded speech from Mr. Sun Zhifang. Discussion and communication at the meeting centered on four major issues which were the theory of quantitative economics, input–output analysis, econometric models and economic forecasting, whose

---

[3] School of Planning and Statistics was established at Renmin University of China in March, 1984. It was renamed as School of Statistics in 2004.

outcome was compiled into a book named *Theory, Model and Forecast of Quantitative Economics* that was published by Energy Publishing[4] in 1983.

From July to August, 1982, the quantitative training class was co-hosted by the China Quantitative Economics Society and the Chinese Academy of Social Sciences with more than 40 students. Higher mathematics, input–output method, econometrics and economic forecasting approaches were the main content of teaching.

At the end of 1982, the Practical Economic Forecast Meeting was co-hosted by the China Quantitative Economics Society and Department of Economic Management at Wuhan University. Conventioneers and staff added up to more than one hundred.

1983

From June 29 to July 6, 1983, the National Seminar of the Application of Input–Output Method was held by the China Quantitative Economics Society and undertaken by the Statistical Bureau of Shanxi Province. About 108 people participated in this meeting and 61 dissertations, reports and other materials were handed in. Discussions and articles at the meeting were collected in the book named *The Application of Input–Output Method in China*, which was published by Shanxi People's Publishing House.

From July 25 to September 5, 1983, the Chinese Association commissioned the Electronic Computing Site of Heilongjiang Province with holding a quantitative economics training class in Harbin. The courses covered the input–output method, econometric models, and economic forecasting. More than 50 trainees attended the class.

On August 23, 1983, the China Quantitative Economics Society had a board meeting in Beijing hosted by Director Wu Jiapei, where 3 proposals were passed.

From October 24, 1983 to January 11, 1984, the study program of quantitative economics was co-hosted by the China Quantitative Economics Society and the computing center at the Ministry of Foreign Economics

---

[4]Energy Publishing was founded in Beijing in November, 1981 and closed in December, 1989, affiliated with Energy Research Institute at National Development and Reform Commission.

and Trade. More than 30 trainees grasped the basic knowledge in terms of input–output method, econometrics and economic forecasting via participating in the program.

1984

From June 11 to 13, 1984, the Second Annual Meeting of the China Quantitative Economics Society was hosted by the Science and Technological Committee of Anhui Province[5] in Hefei, Anhui. More than 150 conventioneers participated in the meeting. Over one hundred dissertations were handed in at the meeting were compiled into 2 books, one of which was named *Research on the Chinese Macro-Economic Model*. Moreover, the China Quantitative Economics Society was renamed as the Chinese Association of Quantitative Economics at the meeting.

On July 10, 1984, the Conference on the Establishment of Liaoning Provincial Quantitative Economics Society and the First Symposium was held in the Benxi Steel Group in Liaoning Province. Mr. Zhou Fang, the secretary-general of the China Quantitative Economics Society, on behalf of the society, attended the meeting to express congratulations and delivered a speech titled *The Object, Content and Method of Quantitative Economics Research*. The meeting invited more than one hundred and twenty conventioneers and received fifty-plus articles and reports.

1985

From September 16–21, 1985, the Chinese Association of Quantitative Economics convened a National Seminar on the Application of Econometric Methods, where 95 people were present and 63 dissertations were handed in. Discussions at the meeting were published in a book named *The Application of Econometric Methods in China* by *Outlook* in China Press.

From October 5 to 10, 1985, the Branch of Colleges and Universities of Chinese Association of Quantitative Economics held the Foundation Meeting and Teaching Experience Seminar in Beidaihe, Hebei Province. The meeting welcomed 55 representatives from 45 comprehensive universities, financial universities and institutes of science and technology that

---

[5]Now the Science and Technology Department of Anhui Province.

are located in 22 provinces, municipalities and autonomous regions. The Chinese Association of Quantitative Economics, Tsinghua University and Renmin University of China congratulated the meeting. It was honored to have at the meeting Professor Ma Bin, who was also the counselor of the Search Centre of Economic, Technological and Social Development at the State Council, Mr. Wu Jiapei, the President of Chinese Association of Quantitative Economics and researcher, Vice Director He Jieren of School of Economics and Management at Tsinghua University, and Mr. Liu Shucheng, the Deputy Director of Institute of Quantitative and Technological Economics at the Chinese Academy of Social Sciences and Secretary General of the Chinese Association of Quantitative Economics. They all made a speech to the conference. The meeting reached a consensus that with the guidance of State Education Commission and Chinese Association of Quantitative Economics and the concerted effort of conventioneers, the academia of quantitative economics at national higher institutions will produce more talented managers for the modern economy.

From October 15–19, 1985, National Enterprises Symposium on Quantitative Economics and Foundation Meeting of Enterprise Branch of the Chinese Association of Quantitative Economics was hosted in Anshan, Liaoning Province. More than 120 conventioneers from 68 companies, scientific research organizations, educational institutions and government offices of 21 provinces, municipalities and autonomous regions were present. Those who were invited to the conference included Mr. Ma Bin, the counselor of the Search Centre of Economics, Mr. Zhao Xinliang, the Vice Director of the Liaoning Provincial Commission of Economic Planning, Mr. Zhang Xianhuan, the Vice Mayor of Anshan, Liaoning Province, Mr. Wu Jiapei, the President of Chinese Association of Quantitative Economics and Researcher, Mr. Liu Shucheng, the Deputy Director of Institute of Quantitative and Technological Economics at the Chinese Academy of Social Sciences and Secretary General of the Chinese Association of Quantitative Economics, and Mr. Li Bingquan from the Institute of Systems Science at the Chinese Academy of Social Sciences. More than 60 dissertations were handed in to the conference. Moreover, both the *Regulations of Enterprise Branch* and *An Open Letter to the Managers of Large and Medium Enterprises in China* were approved and the first council and counselors elected. Upon the finish of the meeting, the first council meeting was held, where the president and vice president were elected, the secretariat was founded and major responsibilities of the branch were initialized.

On November 16, 1985, Lawrence Klein, the well-known economist and econometrician, Nobel Economics Laureate of 1980 was invited to a panel discussion by the Institute of Quantitative and Technological Economics at the Chinese Academy of Social Sciences. He talked about research on quantitative economics and introduced international trends of econometrics.

1986

On February 28, 1986, the enlarged standing committee meeting was held in Beijing by the 2nd council of the Chinese Association of Quantitative Economics. Three crucial matters, including academic issues, work and responsibilities, and personnel management, were discussed at the meeting. Mr. Zhang Shouyi presided over the meeting, where Mr. Wu Jiapei delivered a speech that was titled The Reform of Economic System and Quantitative Economics and on behalf of the association Mr. Liu Shucheng reviewed the work of the organization in the past year and proposed a working plan for 1986.

From November 15–25, 1986, the Academic Seminar on the Application of Input–Output Method was co-hosted at Kunming Institute of Technology[6] by the Enterprise Branch of the Chinese Association of Quantitative Economics and the Department of Management Engineering of the institute. During the Seminar, the Enterprise Branch held its 2nd council meeting at which more than eighty conventioneers from companies, research institutes, colleges and universities of all over China were present.

From November 16–18, 1986, the 2nd council meeting of Enterprise Branch of the Chinese Association of Quantitative Economics was convened. The meeting reported the preparation work of this conference, summarized work from 1985 to 1986, projected the responsibilities of the

---

[6]It is now named as the Kunming University of Science and Technology (KUST). In September 1954, Kunming Institute of Technology (KIT) became an independent public higher education institution. In 1999, the former KIT and former Yunnan Polytechnic University amalgamated into today's KUST.

Branch from 1986 to 1987, and augmented the council by 6 members. The Council considered the branch as a mass academic community and therefore required that every member and representative should insist on a spirit of teamwork and boost up the work of the branch and the research of enterprise quantitative economics by a mutual understanding and a concerted effort.

1987

On January 9, 1987, the Standing Council Meeting of the Chinese Association of Quantitative Economics presided over by the President Mr. Wu Jiapei was held in Beijing. Mr. Ma Bin attended the meeting and delivered a speech. Mr. Liu Shucheng, the Secretary-General of the Chinese Association of Quantitative Economics, Mr. Li Yiyuan, the President of the Branch of Colleges and Universities, and Mr. Wu Liansheng, the Vice President of the Enterprise Branch reviewed the work in 1986 and made plans for 1987. In addition, the standing council decided that collective members and individuals could join any branch that was relevant to them and have a role to play. By the end of 1986, there were almost 1,000 collective members of the association. As basic units of the association, they contributed to the brisk academic activities, as well as the fulfillment of duties and tasks endued to the association.

On March, 1987, the Chinese Input–Output Association, which was subordinate to the Chinese Association of Quantitative Economics, was established by the conjoined efforts of Renmin University of China, Institute of Systems Science at the Chinese Academy of Social Sciences, and National Bureau of Statistics of China.

On September 15, 1987, the Standing Council Meeting of the Chinese Association of Quantitative Economics was held in Beijing. The meeting concerned the coming 3rd annual meeting of the association and suggested that the 3rd council meeting should elect an honorary president and hire several counselors. In addition, it discussed the amendment of regulations of the association and other crucial matters.

In September, 1987, the Enterprise Branch of the Chinese Association of Quantitative Economics convened the 3rd Seminar on the Application of Enterprise Quantitative Economics in Wuxi, Jiangsu Province. This event welcomed 46 attendees and more than 30 articles. Discussions centered on the application of quantitative economic methods such as economic forecast techniques, value engineering, and econometric methods to modern enterprise management.

From October 19 to 24, 1987, the 3rd annual meeting of the Chinese Association of Quantitative Economics was held in Wuhan, Hubei Province. More than two hundred and forty representatives from government offices, research institutes, institutions of higher learning, and enterprises nation-wide participated in this meeting. Nearly two hundred articles about quantitative economic research, application and teaching were handed in. At the meeting that was held on the morning of October 23, the Secretary General of the 2nd council put forward amendment of association regulations and election procedure for the 3rd council in his report. Moreover, the third council was elected with Mr. Zhang Shouyi as the president. The new council, according to the meeting, was supposed to recruit more young members and set up 6 main teams in charge of every domain of work. *Macro-Economic Management and Quantitative Economic Models* published by China Economic Publishing House was the outcome of this meeting.

On December 11, 1987, the Third Standing Council of the Chinese Association of Quantitative Economics held its first meeting hosted by Mr. Zhang Shouyi in Beijing. At the meeting, the primary plans and suggestions on constructing the association were demonstrated by Mr. Shen Lisheng, the secretary general of the association. Operational situations of Input–Output Branch, Enterprise Branch and Branch of Colleges and Universities were reviewed by President Shao Hanqing, Vice President Wu Liansheng and President Li Yiyuan respectively. Moreover, 4 proposals were passed at the meeting in principle. Firstly, 10 young workers who have made a considerable contribution to research, teaching, application promotion or any other work related to the association would be selected and granted Quantitative Economics Award every 3 years. Secondly, a preparatory group made up of Mr. Zhou Fang, Wang Tong, Yuan Fengqi, and Zhong Xueyi would be in charge of building the Branch of Mathematical Economics. Thirdly, the meeting engaged Professor Zhang Qiren, Vice Director of Graduate School of Chinese Academy of Social Sciences, as the Counsellor of the Chinese Association of Quantitative Economics. Fourthly, responsibilities of the association should be taken collectively rather than by the few, and correspondingly adjunct

Association of Quantitative Economics, and Higher Education Institutions Representatives Symposium that were held successively in June, 1990 in Beijing. The conference, presided over by Mr. Li Yiyuan, the President of the branch, discussed and reached a consensus on the branch's future responsibilities and plans. In addition, Mr. Wang Tongsan, the Secretary General of the Chinese Association of Quantitative Economics, attended the meeting.

On October 9, 1990, the Establishment Meeting of Guangdong Provincial Association of Quantitative Economics was held in Guangzhou, Guangdong Province. Professor Wu Jiapei, the honorary President of Chinese Association of Quantitative Economics (CAQE), and Liu Shucheng, the Vice Standing President of CAQE and Vice Director of the Institute of Quantitative and Technical Economics at the Chinese Academy of Social Sciences, attended the meeting to express their congratulations. The conference approved regulations for the new association and formed a council and a standing council with Mr. Li Hongchang as the President, Professor Dou Yingjun, the honorary president, and Mr. Wan Zuoxin, Mr. Wang Rui the vice presidents, and others.

1991

On January 12, 1991, the first meeting of the Fourth Standing Council of Chinese Association of Quantitative Economics was held in Beijing, with Professor Zhang Shouyi, the president of the association as the host. Standing members of the council and councilors that were 24 in total and members of the secretariat attended the meeting. *The Regulations of Chinese Association of Quantitative Economics* that was approved at the conference was published on the second issue in 1991 of *Journal of Quantitative and Technical Economics*.

On January 23, 1991, the second meeting of the Fourth Standing Council of the Liaoning Provincial Association of Quantitative Economics was held in Liaoning Provincial Training Center of Cadre for Economic Planning in Shenyang. About 16 standing members of the council were present at the meeting, and another seven attended the meeting as non-voting delegates. The conference was presided over by President Hua Youtai, who also introduced the main content of the Third Congress of Liaoning Federation of Social Sciences and the Liaoning Symposium of Technical

Economics and Modernized Management. Besides this, Mr. Yu Hongjun of the standing council gave a talk to review the Fourth Annual Meeting of the Chinese Association of Quantitative Economics.

On June 1, 1991, Shaanxi Provincial Association of Quantitative Economics was founded in Xi'an and affiliated to Xi'an University of Finance and Economics. The approvals from the Chinese Association of Quantitative Economics and Shaanxi Commission for Structural Reforms, and a congratulation letter from Professor Zhang Shouyi, the President of CAQE, were read out. The conference passed the regulations for the new association and elected Mr. Li Zhihan as the President, Mr. Li Kewen and Mr. Hu Changzhu as Vice Presidents, and Mr. Hu Changzhu as Adjunct Secretary General.

On July 18, 1991, the Standing Council Meeting of the National Association of Mathematical Economics discussed issues concerning the Mathematical Economics Seminar that was to be held, examined, and approved conference theses, and explored activities for the association in the future.

In July, 1991, the Second National Input–Output Annual Meeting was held in Baotou, Inner Mongolia Autonomous Region. Conference papers were published in *Contemporary Application and Development of Input–Output in China* afterwards.

From October 10 to 14, 1991, the Third Annual Meeting and the Seventh Seminar of The Enterprise Branch of Chinese Association of Quantitative Economics was convened in Dalian, Liaoning Province. More than 90 people who were professors, scholars or senior engineers engaged in enterprise quantitative economics and other economic research participated in the meeting, contributing almost 40 papers. This seminar, presided over by President Yuan Dongzhu, was a grand and comprehensive academic event. In addition, Professor Zhou Fang, the counselor of the Chinese Association of Quantitative Economics, was invited to the conference.

From October 22 to 23, 1991, the First Annual Meeting of the Shandong Provincial Association of Quantitative and Technical Economics was held in Jinan, Shandong Province. More than 140 conventioneers from higher education institutions, enterprises and research institutes in Shandong Province were present. The meeting received 44 papers, as well as the generous support from government offices such as Shandong Federation of Social Science Circles and Shandong Association for Science & Technology. In addition, Professor N.T. Wang (Wang Niantzu), the Senior Researcher of Columbia University, former director of United Nations Center on Transnational Corporation, gave a report of significance.

Notice to Collective Members of Chinese Association of Quantitative Economics was published on the eleventh issue in 1991 of *Journal of Quantitative and Technical Economics*, announcing that any community which was engaged in research, teaching and application of quantitative economics and would abide by the regulations of the association could become a collective member by filling in registration forms, handing in membership dues on time, and going through the examination of the secretariat. At the same time, the notice mentioned that individual membership was not available at that time and councillors should have the organizations or companies that they worked for registered as a collective member.

From November 5 to 7, 1991, the Branch of Mathematical Economics of the Chinese Association of Quantitative Economics (also known as National Association of Mathematical Economics) held its first academic seminar in Beijing since its establishment in 1990. Nearly 40 representatives from higher education institutions and scientific research organizations were present. Twenty plus articles were received. It was an honor to have at the opening ceremony Mr. Wang Tong, the vice director of the State Information Center and honorary president of the National Association of Mathematical Economics, Mr. Zhang Shouyi, the President of the Chinese Association of Quantitative Economics, Mr. Yu Di, the Director of Beijing College of Economics,[10] Mr. Zhou Fang and He Ju Huang, Councillors of the National Association of Mathematical Economics, Mr. Yuan Fengqi, the President of the association, Mr. Zhang Yuansheng, Wei Quanling, ZhongXueyi and Ms. Jin Xianglan, the Vice Presidents, and others.

1992

On January 30, 1992, the second meeting of the Fourth Standing Council of the Chinese Association of Quantitative Economics was held in Beijing. About 26 standing members of the council and counselors, and members

---

[10]The Beijing College of Economics and the Beijing Institute of Finance and Trade merged in June 1995, forming the Capital University of Economics and Business.

of the secretariat were all present. The meeting was presided over by President Zhang Shouyi. Ma Bin, Wu Jiapei and Zhou Fang made speeches at the meeting. According to the reports of men in charge of Input–Output Branch, Branch of Mathematical Economics, Branch of Colleges and Universities, and Enterprise Branch, the 4 branches had completed their plans for academic activities in 1991. Moreover, detailed situations of academic activities, trends of development and plans for the year 1992 were reported. Besides this, the meeting held that there were still a number of people who knew little about quantitative economics within the academia of economics and the issue of whether quantitative economics should be listed as a major for bachelor's or postgraduate degrees remained unsolved.

The *Dispatches of Collective Members of Chinese Association of Quantitative Economics (first issue)* was published in the fourth issue in 1992 of *Journal of Quantitative and Technical Economics*, and so was an article titled "Make a Good Job of Publishing the Dispatches" written by Wu Jiapei and a list of local associations of quantitative economics.

From April 28 to 30, 1992, the second annual academic meeting of Jiangxi Provincial Association of Quantitative Economics was held in Shangrao, Jiangxi Province. More than 70 representatives of members of council and authors of articles handed in were present, of whom 80% and more were scholars of young and middle age. This demonstrated the prospects of a bright future in the field of quantitative economics in Jiangxi Province. Professor Zhang Shouyi, the President of the Chinese Association of Quantitative Economics, and Professor Liu Shucheng, the standing Vice President sent their congratulation letters on behalf of the Chinese Association of Quantitative Economics. The meeting received more than forty academic articles, concerning four aspects: (1) The role quantitative economics had played in Chinese academia and in the reform and opening-up; (2) the setting of standards and indicators to evaluate the vitality of an enterprise; (3) contributing to boost Chinese economy by making best advantages of quantitative economics; (4) the next phase of development in theory and application research of quantitative economics, including theoretical innovation, popularization of quantitative economical methods of high practical value, and the extended application of development and methods of quantitative economical software.

From August 7 to 13, 1992, the second academic seminar of the Branch of Mathematical Economics was held in Chengdu, Sichuan Province. Nearly 30 representatives from higher education institutions and research organizations from all over China attended the meeting and had their 33 papers announced. The seminar was hosted and supported by Chengdu University of Science and Technology.[11] The seminar welcomed He Juhuang, the counselor of the branch, President Yuan Fengqi, Vice President Jin Xianglan and members of the council and representatives from other organizations. The part the young and middle-aged representatives played in the proceedings at this seminar showed an obvious increase compared to the last one.

From October 12 to 16, 1992, the eighth academic seminar and council meeting of the special committee on enterprise of the Chinese Association of Quantitative Economics was held in Xiamen, Fujian Province. Conventioneers were dominated by scholars and representatives of enterprises engaged in research and the application of enterprise quantitative economics from 14 provinces, as well as members of the council, accounting for almost 60. In addition, 35 articles were handed in to conference. At the meeting, Vice President Wu Liansheng hosted the opening ceremony and Zhang Shouyi, President of the Chinese Association of Quantitative Economics announced his report titled *Several Issues on Economic Development and Reform*, which was a great interest to the conventioneers. Moreover, research achievements were reflected through team discussions and exchange of ideas during the seminar.

1993

On February 18, 1993, the third meeting of the fourth standing committee of the Chinese Association of Quantitative Economics was held in Beijing. Members of the Secretariat were present. About 22 members of council and counselors in Beijing attended the meeting. President Zhang Shouyi hosted the meeting. Mr. Li Jingwen, Zhou Fang, Chen Xikang, Li Zegao, and Liu Shucheng made speeches. Mr. Li Yiyuan, Yuan Fengqi, Wu Liansheng, and Liu Qiyun reported separately work that had been done by 4 branches

---

[11]Chengdu University of Science and Technology was founded in 1952 and merged into Sichuan University in 1993.

(of colleges and universities, of mathematical economics, of enterprise, and of input–output). In addition, 8 excellent staff of quantitative economics (1990–1993) were elected through an anonymous vote.

From September 21 to 26, 1993, the fifth annual meeting of the Chinese Association of Quantitative Economics was co-hosted by the association and Shandong Association of Quantitative and Technical Economics in Jinan, Shandong Province. About 133 people from central ministries, scientific research organizations, higher education institutions and enterprises attended the meeting. Professor Zhang Shouyi, the President of the Chinese Association of Quantitative Economics presided over the meeting, where Professor Hu Jijian, the president of Shandong College of Economics,[12] delivered a welcome speech, and Professor Liu Shucheng, the standing Vice President of Chinese Association of Quantitative Economics, announced the congratulation letter from Professor Liu Guoguang, the Vice Director of Chinese Academy of Social Sciences. Moreover, the meeting conferred certificates of honor and trophies on the 8 elected excellent staff of quantitative economics (1990–1993), namely Wang Shizhou, Li Zinai, Xie Fang, Zhang Mingxing, Huang Dengshi, Cheng Jianlin, Wang Shengling and Tao Lixin. Besides this, the meeting received 94 academic papers. Scholars and representatives exchanged ideas in a serious, friendly and passionate manner during team discussion and found the discussion beneficial. The meeting also elected the fifth council of the Chinese Association of Quantitative Economics, which was stationed in seven regions throughout China, namely Beijing, northeast China, north China, south China, south-central China, southwest China, and northwest China. Of the 76 members, there were 15 newly-elected members of the council, making up 20% of the whole. The first meeting of the new council was held soon afterwards, deciding by vote the fifth standing council of the association. Professor Wu Jiapei was unanimously elected as the honorary president, Professor Zhang Shouyi, the President and Professor Liu Shucheng, the Vice Standing President. The meeting elected 10 Vice Presidents and 22 standing memebers of the council, of which 11 were elected as a standing memeber for the first time. The president of the

---

[12]Shandong College of Economics (originally known as Shandong College of Finance and Economics), founded in 1952, was emerged into Shandong University of Finance and Economics together with Shandong Finance Institute in 2012.

association appointed Mr. Zhang Jingzeng as Secretary-General, and Mr. Li Fuqiang and others as Vice Secretary-Generals to lead the secretariat. And 20 well-known economists were employed as Counsellors.

In September, 1993, the Enterprise Branch of the Chinese Association of Quantitative Economics held its ninth academic seminar in Urumqi, Xinjiang Uygur Autonomous Region, focusing on decision optimization and the application of computer-aided decision-making. More than 60 conventioneers were present and their papers discussed at the meeting.

1994

In 1994, the School of Economics and Management at Tsinghua University was the first in China to set up a master's pilot for a major in quantitative economics.

The third and sixth issue in 1994 of *Journal of Quantitative and Technical Economics* introduced those who were awarded as excellent staff of quantitative economics (1990–1993), including Wang Shizhou, Li Zinai, Xie Fang, Zhang Mingxing, Huang Dengshi, Cheng Jianlin, Wang Shengling, and Tao Lixin.

On June 26, 1994, Zhejiang Association of Quantitative Economics held its founding meeting at Zhejiang University of Technology in Hangzhou, Zhejiang Province. It discussed and passed the regulations and produced the first council with Wu Tianzu as the President, Zhang Mingliang, Yuan Fei as Vice President, Dong Taiheng as Secretary-general and Luo Guoxun as Vice Secretary-General.

From August 12 to 15, 1994, the third annual meeting of Chinese Input–Output Association was held in Yinchuan, Ningxia Hui Autonomous Region. About 48 representatives from research organizations, higher education institutions, government sectors and enterprises handed in 39 papers. At the opening ceremony, honorary President Professor Chen Xikang made the opening speech. Vice President Professor Liu Qiyun reported work that the association had done and Vice President and Secretary-General Qi Shaocheng updated the progress on bidding for the first class association. At the closing ceremony, Vice President Professor Li Zinai brought up working plan for the next year. In addition, papers collected at the meeting were published in *Empirical and Innovative Study on Contemporary Chinese Input and Output*.

From September 19 to 23, 1994, the fourth annual meeting of The Enterprise Branch of Chinese Association of Quantitative Economics and tenth academic seminar was held in Xinyu, Jiangxi Province. Zhang Shouyi, the President of Chinese Association of Quantitative Economics and Liu Shucheng, Vice Standing President sent their congratulation letters. President Yuan Dongzhu gave an inaugural address, summarizing work the council had done for the past 3 years and pointing out the future direction of effort. The council was reelected. Yuan Dongzhu became the new president, Wang Shizhou the Vice Standing President and secretary general, and Wu Liansheng and Li Fuqiang became the vice presidents. On the new council meeting, Wang Shizhou outlined the work to do for the next 3 years and explicated the work plan for 1995.

1995

On January 17, 1995, the fifth standing council of the Chinese Association of Quantitative Economics held its second meeting in Beijing. Standing members of the council in Beijing and counselors, numbering 26 people, were present, as well as members of the secretariat as non-voting delegates. President Zhang Shouyi presided over the meeting and made a speech. Respectively, on behalf of the secretariat, the Professional Committee of Mathematical Economics, of Input and output, of Colleges and Universities, and of Enterprises, Zhang Jingzeng, Yuan Fengqi, Tong Rencheng, Li Yiyuan and Wu Liansheng made their work reports. In addition, the meeting showed appreciation for what the association had done in 1994 and spoke highly of the contribution the association, and the Professional Committees in particular, had made to the development of the Chinese quantitative economics.

From May 13 to 16, 1995, the Shanghai Economist Association and Zhejiang Association of Quantitative Economics co-hosted a free paper session in Mogan Moutain, Zhejiang Province. The meeting was presided over by Professor Hu Zuguang, the Vice President of Zhejiang Association of Quantitative Economics and Deputy Director of Hangzhou College of Commerce. Fifty scholars from Shanghai and Zhejiang attended the meeting, so did Zhang Jingzeng, the Secretary-General of the Chinese Association of Quantitative Economics. Discussions were fruitful, producing more than 40 papers.

From July 21 to 25, 1995, the National Association of Mathematical Economics held its fourth academic seminar in Mount Lu, Jiangxi Province. Professor Zhang Shouyi, the President of the Chinese Association of Quantitative Economics, sent his congratulation letter. Wang Tong, the director of Institute of Economic System and Management at National Committee for Economic System Reform and honorary President of National Association of Mathematical Economics, attended the meeting where he made an important speech. Professor Yuan Fengqi, the President of National Association of Mathematical Economics, gave the opening speech and presided over the meeting. At the closing ceremony, Associate Professor Li Zijiang, also the Vice President of National Association of Mathematical Economics, made a summary of the meeting. Besides this, the meeting received 43 attendees and more than 30 papers.

From July 28 to 30, 1995, the Quantitative Economics and Economic Discipline Construction Symposium was organized and hosted in Beijing by Branch of Colleges and Universities of the Chinese Association of Quantitative Economics. About 58 representatives from 34 higher education institutions and research organizations such as the Institute of Quantitative and Technical Economics at the Chinese Academy of Social Sciences attended the meeting. More than 20 academic papers were handed in. Conventioneers affirmed the necessity and significance of holding a meeting like this at the time of the decennial of the founding of the Branch. In light of a decade of prosperous development in quantitative economics, the meeting offered an opportunity to discuss and probe into the subject nature of quantitative economics, its development direction and the role it has played in the socialist market economy and economic discipline construction. The results of the meeting would surely cast a profound influence on the development of Chinese quantitative economics research and economy related subjects.

From October 12 to 14, 1995, the Enterprise Branch of Chinese Association of Quantitative Economics celebrated the tenth anniversary of its establishment and held the eleventh academic symposium in Benxi, Liaoniang Province. Yuan Dongzhu, the President of the branch association and Chief Economic Manager of Anshan Iron and Steel Company, presided over the meeting and delivered the opening speech. Vice President Wu Liansheng made a summary of the meeting. Conventioneers from enterprises, scientific research institutions, colleges and universities, and government offices accounted for more than 40 attendees. Fourteen papers were handed in to the meeting.

The eleventh issue in 1995 of *Journal of Quantitative and Technical Economics* published "Selected Theses on the Forth Academic Seminar of National Association of Mathematical Economics". About 24 papers were introduced, including "Principle of General Market Equilibrium" written by Professor Wang Tong.

1996

On January 18, 1996, the fifth standing council meeting of the Chinese Association of Quantitative Economics was held in the Institute of Quantitative and Technical Economics at the Chinese Academy of Social Sciences. The meeting, presided over by President Zhang Shouyi, welcomed 40 standing members of the council in Beijing and counselors, and members of secretariat as nonvoting delegates. It summarized the association's work in 1995 and made suggestions on work to do in 1996. Respectively on behalf of the Professional Committee of Mathematical Economics, of Enterprises, and of Input and output, Yuan Fengqi, Wu Liansheng, and Liu Qiyun gave an account of what their Professional Committee had done in the past years. Representatives also put forward five suggestions as to future tasks for the association: Firstly, pay attention to subject construction of quantitative economics; secondly, make efforts in basic quantitative economics education; thirdly, continue the research on the application of quantitative economics to the real economy; fourthly, build awareness of service to policy-makers; and lastly have research and application results of quantitative economics published in economic periodicals.

The first issue in 1996 of *Journal of Quantitative and Technical Economics* published several articles about theory of quantitative economics, including "Strengthen Theoretical Research to Develop Quantitative Economics" written by Wu Jiapei, "Quantitative Economics: A Major Force in the Construction of Socialist Marketing Economics" by Li Yiyuan, and "Keep Up with the Main Trend, Be Among the First-Rate: A Historical Mission for Chinese Quantitative Economics" by Li Zinai.

From September 23 to 26, 1996, to popularize basic knowledge about economic game theory and promote its development in China, the Chinese

Association of Quantitative Economics sponsored the National Economic Game Theory Symposium in Hangzhou, Zhejiang Province. The symposium was co-hosted by the Zhejiang Association of Quantitative Economics, Shanghai Association of Quantitative Economics and Hangzhou College of Commerce. President Zhang Shouyi presided over the seminar and made a speech. More than 60 scholars and specialists from all over China attended the meeting, as well as Huang Shuyan, the counsellor of the Chinese Association of Quantitative Economics and President of the Shanghai Association of Quantitative Economics, Hu Zuguang, the Vice President of the Chinese Association of Quantitative Economics and President of Hangzhou College of Commerce, Feng Wenquan, the counsellor of the Chinese Association of Quantitative Economics, Liu Qiyun, the Vice President of the Chinese Association of Quantitative Economics, Zhang Jingzeng, the standing member of the council of the Chinese Association of Quantitative Economics and Secretary-General, Dong Taiheng, the Member of the council of the Chinese Association of Quantitative Economics and Secretary-General of Zhejiang Association of Quantitative Economics, and He Lunzhi, Luo Ronggui and Wang Fuxin, the members of the Chinese Association of Quantitative Economics. The meeting invited Ke Wei, Li Zijiang and Wang Guocheng to deliver plenary speeches.

1997

From August 27 to 29, 1997, the sixth annual meeting of Liaoning Association of Quantitative Economics was held in Nandaihe, Hebei Province. Jiang Jianli, the Deputy Director of Liaoning Province Information Center, presided over the meeting. The secretariat of the Chinese Association of Quantitative Economics sent a congratulation letter. Ten academic papers were read out for discussion. Besides, Jiang Jianli was elected as the president of Liaoning Association of Quantitative Economics and Tai Tianyu the Secretary-General.

In July 1997, the Professional Committee of Economic Game Theory of the Chinese Association of Quantitative Economics organized the textbook *Modern Economic Game Theory* compilation workshop in Beijing. Professor Zhang Shouyi presided over the workshop and he was the editor-in-chief of this textbook. Main authors Li Zijian, Wang Wenju and Wang Guocheng explicated their contributions to the book.

In August 1997, the sixth annual meeting of the Chinese Association of Quantitative Economics and the fourth annual meeting of the Chinese Input–Output Association (CIOA) were held in Changchun, Jilin Province. Professor K.R. Polenske, the President of International Input–Output Association attended the meeting and all academic exchanges. Discussion and papers from the two meetings made possible the publication of *Introducing Quantitative Economics* by Social Sciences Academic Press and of *Practice and Research of Contemporary Chinese Input–Output* by CIOA.

1998

From January 4 to 7, 1998, the Frontier Financial Theory and Practice Symposium was held in Zhongshan, Guangdong Province. It was originally initiated by Tsinghua University and other 17 universities and hosted by the Finance Research Department at the Chinese Association of Quantitative Economics and the National Center of Economic Research at Tsinghua University. About 65 representatives attended the meeting. More than 40 papers were handed in.

On January 20, 1998, the first executive meeting of the sixth council of the Chinese Association of Quantitative Economics, presided over by President Zhang Shouyi, was held in Beijing. About 28 counsellors and standing members of the council in Beijing attended the meeting. Secretary-General Zhang Jingzeng informed the meeting about the sixth annual meeting that had been hosted in Changchun, Liaoning Province. Those in charge of the Professional Committee of Input and Output, of Colleges and Universities, of Enterprises, and of economic game theory reported their work in 1997 and proposed work plans for 1998. The Vice President, Li Yiyuan, introduced the preparatory work for the establishment of the Professional Committee of Distance Education. Moreover, the meeting reached a consensus as to the founding of the Professional Committee of Distance Education and the Finance Research Department. The Professional Committee of Distance Education was to be subordinated to China Central Radio & TV University. Sun Tianzheng was appointed as the Director of the committee and at the same time a member of standing council of the Chinese Association of Quantitative Economics. The Finance Research Department was to be subordinated to Tsinghua University,

From November 17 to 19, 2001, the Shanghai Association of Quantitative Economics and Zhejiang Association of Quantitative Economics hosted an academic seminar in Tianmu Mount in Lin'an, Zhejiang Province. Representatives from 16 colleges and universities in Shanghai and Zhejiang Province and from scientific research organizations attended the meeting, handing in 20 plus articles. Besides this, Zhejiang Association of Quantitative Economics held the second membership congress when council of the second term was elected.

2002

From May 29 to 31, 2002, the Chinese Association of Quantitative Economics and Huaqiao University co-hosted the 2002 Annual Conference of Chinese Association of Quantitative Economics in Chen Jiageng Memorial Hall of Huaqiao University, Quanzhou, Fujian Province. Wu Jiapei, the honorary President of the Chinese Association of Quantitative Economics and honorary Director of the School of Economics and Management at Huaqiao University, hosted the opening ceremony. Professor Wang Tongsan, the President of the Association and Director of the Institute of Quantitative and Technical Economics at the Chinese Academy of Social Sciences, delivered a speech. Wu Chengye, the President of Huaqiao University, gave a welcoming address. At the opening ceremony, Daniel McFadden, the 2000 Nobel Prize winner in Economics, Professor Wang Tongsan, Wu Jiapei, Zhang Shouyi and Wang Tong gave special reports respectively. Moreover, during team discussions, topics like quantitative analysis of finance and capital, quantitative analysis of regional economics, theory and application of quantitative economics, and integration of theory and application of quantitative economics were deliberated. In addition, the council meeting of the association was held during the conference, discussing the responsibilities of the council and supplementing members to the council. In addition, more than 160 representatives were present, contributing to conference articles of more than 100.

From August 15 to 17, 2002, the thirteenth academic seminar of the Enterprise Branch and Annual Conference of Zhejiang Association of Quantitative Economics were convened in Ningbo, Zhejiang Province. More than 40 conventioneers from all over the country attended the meeting. Zhang Xukun, the President of the Enterprise Branch, Professor of Zhejiang

University, and Director of the School of Commerce at Ningbo University, and Wu Liansheng, the Vice President of the Enterprise Branch, hosted the meeting. In addition, President Zhang Xukun suggested that the Enterprise Branch should attract more people from companies on the one hand to cater to the needs of the enterprise and on the other hand to liven the academic environment and enrich exchanges via a combination of theory with practice.

2003

On January 15, 2003, the fourth meeting of the seventh term of standing council of the Chinese Association of Quantitative Economics was held in Beijing, inviting almost 30 standing members of the council from all over China to present. It was hosted by Li Fuqiang, the Vice Standing President. President Wang Tongsan made a speech and the secretariat summarized their work and final report of 2002 and proposed financial budget for 2003. Zhang Jingzeng, the Vice President and adjunct Secretary-General, explicated the new term of the council and issues concerning fees of individual and collective membership.

From December 3 to 5, 2003, the eighth membership congress and academic seminar of the Chinese Association of Quantitative Economics was held in Hunan University. At the conference, the seventh council gave a working report and the eighth council and academic committee were elected and formed. The meeting also discussed and examined a financial report that was handed in. At the same time, the 2003 annual meeting of the association with the theme of "comprehensive well-being, quantitative economy, and credit management" was hosted. The meeting received 127 academic papers, 76 of which were discussed. James Heckman, the Nobel Prize winner in Economics and Professor of Chicago University, attended the meeting and gave a speech, and so did Professor Wu Jiapei, honorary President of the association, and Professor Li Zinai from Tsinghua University, who was also the standing Vice Director of the academic committee. In addition, more than 220 representatives, among whom 70 were professors, were present. Their discussion covered many aspects centering on the theme of the meeting, such as evaluative indicators to build a well-off society in an all-around way, credit management, economic growth and scientific development, finance and monetary policy, regional economic development, industrial economy evaluation, theory and methodology of quantitative economics and other practical issues.

2004

From May 29 to 31, 2004, the annual meeting of the Chinese Association of Quantitative Economics was held in Southwest Jiaotong University, welcoming 234 representatives from all over the country. Centering on the theme of "quantitative economics and coordinated development", 228 articles were handed in to the conference. Therefore, team discussions covered issues on econometrics, mathematical economics, economic game theory, financial and trade investment, quantitative finance, harmonious development and enterprise innovation. Moreover, Robert Mundell, the 1999 Nobel Prize winner in Economics, attended the meeting and addressed it with a speech. And the council meeting of the Chinese Association of Quantitative Economics was also hosted during the conference.

From June 10 to 13, 2004, the fourth academic symposium of the Enterprise Branch and the Innovative Private Economy Management Forum were convened at the School of Economy, Wenzhou University, Zhejiang Province. Forty representatives from all over China were present. The conference was presided over by Zhang Xukun, the President of Enterprise Branch, and Wu Liansheng, the Vice President. Professor Gregory Chow from Princeton University, and Wang Tongsan, the President of the Chinese Association of Quantitative Economics made speeches at the conference. Besides, representatives discussed innovative management in private enterprises, which turned out to be fruitful.

During August 2004, the sixth annual meeting of Chinese Input–Output Association was held in Kunming, Yunnan Province. Conference proceedings titled *Input–Output Theory and Practice in China (2004)* was published after the meeting.

2005

On January 22, 2005, the standing council meeting of the Chinese Association of Quantitative Economics was held in Beijing and presided over by Li Fuqiang, the Standing Vice President. President Wang Tongsan summarized work in 2004 and proposed a working plan for 2005. The meeting passed the final financial accounts of 2004 and budget in 2005.

Representatives decided the theme of 2005 annual meeting and topics for discussion. And Li Zinai, the Vice Director of academic committee, introduced academic activities that had been held in 2004.

From June 4 to 5, 2005, the annual conference of the Chinese Association of Quantitative Economics with the theme of "scientific outlook of development and Chinese economic development" was held in Nanjing University of Finance, receiving over 300 people and 267 articles. Professor Clive Granger, the 2003 Nobel Prize winner in Economics, was invited to the meeting to give a speech. Professor Zhang Xiaotong also addressed a report titled *New Progress in Unit Root Test*. Representatives also participated in the team discussions which were held in 12 different parallel sessions. A council meeting was hosted during the meeting where personnel adjustments were made.

In June 2005, the Fifteenth International Conference of Input–Output Technology was co-organized in Renmin University of China by the International Input–Output Association, the Chinese Input–Output Association, Renming University of China, the National Bureau of Statistics, and the Development Research Center of the State Council. Professors at colleges and universities, scholars in research organizations and government officials from forty countries attended the conference.

In August 2005, the First National Experimental Economics Meeting was co-held in Beijing by China Game Theory Research and Beijing Information Science and Technology University.

From August 23 to 25, 2005, the College of Economics in Shanghai University of Finance and Economics organized the National Advanced Seminar on Quantitative Economics, attracting over 50 scholars and specialists from 30 universities and scientific research centers at home and abroad. Representatives exchanged ideas about the application of micro- and macro-econometrics, econometric analysis and application of financial time series, and the application of theory and method of econometrics and handed in nearly 30 theses.

In December 2005, joining hands with Capital University of Economics and Business, the China Game Theory Research held National Symposium on Game Theory and Experimental Economics in Beijing. At the meeting, Professor Zhang Shouyi suggested that the China Game Theory Research changed its name to China Association of Game Theory and Experimental

Economics, which got the support of all representatives. Ge Xinquan, Li Jianbiao and Fei Fangyu were supplemented as Vice Presidents.

2006

From May 12 to 14, 2006, the annual meeting of the Chinese Association of Quantitative Economics was held in Zhejiang Gongshang University with the theme of quantitative economics and constructing harmonious society. More than 400 representatives from colleges and universities, research organizations, America, Japan and Taiwan, China attended the meeting, handing in over 390 theses. During the meeting, representatives were divided into 20 groups to discuss the theory and application of econometrics, theory and application of economic growth theory, income, distribution, fairness and harmonious society, programming theory, game theory and experimental economics, economic control, system optimization and information theory, and financial econometrics and risk management. In addition, 8 unit members, namely the School of Commerce at Jilin University, Shandong Economic University, Huaqiao University, Chongqing Technology and Business University, School of Economics and Management at Hunan University, School of Economics and Management at Southwest Jiaotong University, School of Economics and Finance at Nanjing University of Finance, and Zhejiang Gongshang University, were awarded Outstanding Contribution Award by the Chinese Association of Quantitative Economics. A Special Academic Work Prize went to Professor Cheng Hsiao from University of Southern California, Professor Koufu Morimune from Kyoto University, Professor Lanh T. Tran from Indiana University, Professor Guan Zhongmin from Academic Sinica in Taiwan, and Professor Wu Bolin from National Chengchi University. Moreover, the meeting produced the council of the ninth term, electing Wang Tongsan as the President, Qi Jianguo and other two as standing Vice Presidents, Wang Naijing and other 14 as Vice Presidents, Wang Libin and other 50 as standing members of the council, and Ma Shu and other 70 as members of the council. In addition, Li Jinhua was nominated and elected as standing Vice President and Adjunct Secretary-General, Wu Jiapei and Zhang Shouyi as honorary Presidents, and Yu Jingyuan and other 26 people as Counsellors.

In September 2006, the Chinese Input–Output Association held the 2006 Annual Input–Output Seminar in Beijing. Its discussion centered on the analysis and application of *2002 Chinese Input–Output Graph*. Professor Masahiko Shimizu from Kelo University and Wang Zaizhe from Rissho University, Professor Chen Xikang and Liu Qiyun were invited to the meeting. The conference also welcomed conventioneers form scholars and graduate students from universities and research organizations such as the National Bureau of Statistics, Renmin University of China, Academy of Mathematics and Systems Science at the Chinese Academy of Sciences, School of Management at the Chinese Academy of Sciences, Tsinghua University, State Information Center, Academy of Macroeconomic Research of NDRC, and Beijing University of Posts and Telecommunications.

In October 2006, the China Association of Game Theory and Experimental Economics hosted an academic annual meeting with the theme of experimental economics, harmony and honesty. Doctor Robert Veszteg from Navarra University and Professor Yan Huibin from California University were invited to the meeting to address reports. Tens of specialists and scholars from all over China discussed and exchanged ideas during the meeting. Research staff from other nations including Argentina and Singapore handed in theses to the meeting.

2007

From May 19 to 20, 2007, the 2007 Annual Meeting of the Chinese Association of Quantitative Economics was held at Wuhan University of Technology, receiving nearly 400 scholars from home and abroad. The theme of this meeting was the theory, method and application of quantitative economics. Clive Granger, the Nobel Prize winner in economics was invited to the meeting to give a keynote speech. Articles that were contributed by teachers and students from universities in China and scholars outside amounted to nearly 300, touching upon every aspect in quantitative economic studies. They were written in not only Chinese but also English, reflecting the openness of the annual meeting and the prestige the association had built in academia. The council decided to issue the Outstanding Contribution Award to Wuhan University of Technology.

In August 2007, China Association of Game Theory and Experimental Economics sent some staff and members to attend the Asia-Pacific ESA Conference in Shanghai. Members of the Research read out their articles for further discussion.

In August 2007, the Seventh Annual Conference of the Chinese Input–Output Association was convened in Nanjing, Jiangsu Province. Conference proceedings named *Input–Output Theory and Practice in China (2007)* were published after the meeting.

From November 24 to 25, 2007, the Second National Advanced Seminar on Quantitative Economics was held in Shanghai Academy of Social Sciences and Fudan University, which was sponsored by Research Center of Econometrics at the Shanghai Academy of Social Sciences, Research Center for Game and Quantitative Economics at Fudan University, Chinese Association of Quantitative Economics, and Shanghai Association of Quantitative Economics and co-organized by the former two organizations. Over 150 professors and scholars from 50 universities, research centers and government offices, including academic leaders, and experts and scholars on the front line, attended the meeting and some of them gave keynote speeches and academic reports. More than 60 articles were handed in to the conference, covering the main themes such as application of macro and micro econometrics, the analysis and application of financial line series, game theory and macroeconomic forecast, and research on theory and methodology of econometrics.

2008

On January 12, 2008, the second standing council meeting of the ninth council of Chinese Association of Quantitative Economics was convened in Beijing and presided over by Vice President Li Fuqiang. Nearly 50 people including counselors, Standing Members of the council and Vice Presidents attended the conference. The secretariat informed standing members of the council of preparatory work for the 2008 annual meeting. It was decided that the 2008 annual meeting would be organized by Ningbo University. The meeting also passed the proposal of setting up the Excellent Paper Award, an expert rostrum, and a Ph.D. forum. In addition, the secretariat reported work that had been done in 2007 and suggested that modern methods should be made use of to popularize the association and to offer better service to members.

From May 16 to 18, 2008, the 2008 Annual Meeting of Chinese Association of Quantitative Economics was convened in Ningbo University, receiving 418 members and representatives and 324 articles. Robert Mundell, the Nobel Prize winner in Economics, attended the meeting and made a keynote speech. Tang Shaoxiang, the Vice President of Ningbo University and Vice Presidents of the association presided over the opening ceremony and Professor Li Fuqiang, the standing Vice President of the association, the closing ceremony. Professor Wang Tongsan, the President of the association and committee member of the Chinese Academy of Social Sciences, addressed both the opening and closing ceremonies. Based on the topics of submitted articles, representatives were divided into eight groups to discuss separately on the theory and method of econometrics, mathematical economics and other sub-disciplines, macroeconomic growth and development, finance, bank, monetary, investment and trade, regional economy, harmonious development, and enterprise and industry economy, and game theory and experimental economics. More than 60 scholars introduced their latest research results during panel discussion. Moreover, according to discussion and regulations of the association, 31 young scholars or graduates were issued the Excellent Paper Award, including 5 first prizes and 26 second prizes. This was the first time that such award had been issued. During the conference, the second plenary meeting of the ninth council of the Chinese Association of Quantitative Economics was hosted.

In June 2008, the China Association of Game Theory and Experimental Economics invited Professor D. Houser from ICES at George Mason University to open an advanced seminar on experimental economics. He lectured on the principles, methodology and application of experimental economics and took charge of field experiments, from which scholars who attended the class benefited a lot.

In June 2008, the Chinese Input–Output Association convened the 2008 Input–Output Symposium in Beijing, discussing the application of fixed-price input–output sequence table from 1987 to 2005. Scholars and graduate students from the State Bureau of Statistics, Renmin University of China, Academy of Mathematics and Systems Science at the Chinese

Academy of Sciences, School of Management at the Chinese Academy of Sciences, Tsinghua University, Tianjin University of Finance and Economics, and Yanshan University attended the symposium.

On November 30, 2008, the third standing council meeting of the ninth council of the Chinese Association of Quantitative Economics was convened in Beijing, receiving more than 50 representatives. Li Fuqiang, the standing Vice President, presided over the meeting. The meeting decided the theme for the 2009 annual meeting and passed the 2008 financial final accounts and budget for 2009. In the end, President Wang Tongsan made a summary speech.

2009

From March 28 to 29, 2009, the 2009 Annual Meeting of Chinese Association of Quantitative Economics was held in Shenzhen University, welcoming more than 600 scholars and representatives from almost 100 universities and research centers. President Wang Tongsan, Tangjie, the Deputy Mayor of Shenzhen, and Zhang Bigong, the President of Shenzhen University attended the opening ceremony. More than 540 academic articles were submitted to the conference. And Guan Zhongmin, Hong Yongmiao, and Lu Maozu gave reports. Conventioneers were divided into 14 groups to discuss 10 topics, including theory and method of quantitative economics, macroeconomic growth and development, currency, bank, finance and capital market, finance and taxation, trade and investment, and harmonious development in regional economies. Moreover, the meeting issued the Excellent Paper Award including 20 first prizes, 22 second prizes and 10 third prizes. In addition, the conference produced the tenth council of the Chinese Association of Quantitative Economics. Professor Wang Tongsan was elected as the President, Li Xuesong from the Chinese Academy of Social Sciences, Hong Yongmiao from Xiamen University, and Xu Xiaoguang from Shenzhen University as Vice Presidents, and Li Jinhua as the Secretary-General.

On May 26, 2009, sponsored by the Chinese Association of Quantitative Economics and organized by Beijing Information Science and Technology University (BISTU), the 30th Anniversary Commemoration of the Establishment of the Chinese Association of Quantitative Economics was held in Beijing. Honorary President Zhang Shouyi, President

Wang Tongsan, Han Qiushi, the Vice President of BISTU, Professor Ge Xinquan, the Director of the School of Economics and Management at BISTU and another 40 scholars and experts attended the meeting. In addition, the association issued to honorary President Wu Jiapei and Zhang Shouyi, the Outstanding Contribution Award in appreciation for their contributions to the establishment and development of quantitative economics in China.

The Brief History of the Development of the Professional Committee of Game Theory

Wang Guocheng

In the early 1990s, the Chinese Association of Quantitative Economics has made game theory a major research direction for some time in the future. Professor Zhang Shouyi, the then President of the association was a Ph.D. supervisor, recruiting and training the first batch of doctors of economics that were engaged in game theory and its application to the Chinese economy.

Under the background of deepening reform in the orientation towards a market economy and three 1994 Nobel Prizes in Economics going to game theory experts, the First National Seminar on Economic Game Theory was held in Hangzhou in September 1996. This conference was sponsored by the Chinese Association of Quantitative Economics and organized by Zhejiang Gongshang University. It welcomed nearly 100 experts and scholars from universities and research centers, and invited You Kewei, Li Zijiang and Wang Guocheng to give keynote speeches. Professor Zhang Shouyi suggested that Wang Guocheng took in charge of setting up the Professional Committee of Economic Game Theory, which won the consent of members of the council and representative at the conference and later the approval of the standing council meeting.

In July 1997, the Professional Committee of Economic Game Theory (in preparation) organized the workshop for editing a textbook named *Modern Economic Game Theory*. Professor Zhang Shouyi was the editor-in-chief, presiding over the workshop. Li Zijiang, Wang Wenju, and Wang

Guocheng were 3 main authors. They talked about their parts of writing separately at the workshop.

Upon the approval by the standing council of the Chinese Association of Quantitative Economics (and by the Ministry of Foreign Affairs), the Establishment Conference and Academic Seminar of the Professional Committee of Quantitative Economics (China Game Theory Research) was held in Beijing from July 17–20, 1998. Leaders of the Chinese Association of Quantitative Economics and men in charge of committees came to offer their congratulations. About 56 scholars and specialists from nearly 40 universities and research centers attended the conference. Moreover, the meeting produced the first council made up of 30 members. Wang Guo was elected as the President, Li Xuesong the standing Vice President, and Wang Wenju, Huang Tao, Chen Xuebin and Li Zijiang as Vice Presidents. Wang Wenju was also employed as adjunct Secretary-General, Li Tao the Vice Secretary-General, and 7 experts including Zhang Weiying, Wang Shouyang as the counsellors. During the conference, conventioneers exchanged their recent research results and discussed about the textbook *Modern Economic Game Theory*. Huang Tao, Wang Wenju and Wang Guocheng introduced their parts of writing respectively.

During the summer vacation in 1999, China Game Theory Research, together with the National Economic Mathematics Committee, hosted the National Seminar on Game Theory and Economic Mathematics at Huaqiao University in Quanzhou, Fujian Province.

In December 2000, China Game Theory Research and the Economic Science Lab at Renmin University of China hosted the workshop of Game Theory and SWARM application, aimed at exploring and promoting game theory and its quantified research and application.

During the summer holiday in 2001, China Game Theory Research sponsored and the School of Economics and Management at Nanchang University organized the National Annual Conference of Game Theory. Academic exchanges went smoothly as expected. In addition, based on needs at work and suggestions from representatives, Gan Xiaoqing and Xie Shiyu were supplemented as Vice Presidents.

From 2002 to 2004, the annual meetings of each Professional Committee were integrated into the annual conference of the Chinese Association of Quantitative Economics.

In August 2005, China Game Theory Research and Beijing Information Science and Technology University co-organized the First National Conference on Experimental Economics in Beijing.

In December 2005, China Game Theory Research and Capital University of Economics and Business co-hosted the National Symposium on Game Theory and Experimental Economics. At the meeting, Professor Zhang Shouyi suggested that China Game Theory Research change its name to China Association of Game Theory and Experimental Economics, which got the support of all representatives. Ge Xinquan, Li Jianbiao and Fei Fangyu were supplemented as Vice Presidents.

In October 2006, China Association of Game Theory and Experimental Economics hosted an academic annual meeting with the theme of experimental economics, harmony and honesty. Doctor Robert Veszteg from Navarra University and Professor Yan Huibin from California University were invited to the meeting to present reports. Tens of specialists and scholars from all over China discussed and exchanged ideas during the meeting. Research staff from other nations including Argentina and Singapore handed in their theses to the meeting.

In August 2007, China Association of Game Theory and Experimental Economics sent some staff and members to attend the Asia-Pacific ESA Conference in Shanghai. Members of the Research read out their articles for further discussion.

In June 2008, China Association of Game Theory and Experimental Economics invited Professor D. Houser from ICES at George Mason University to open an advanced seminar on experimental economics. He lectured on the principles, methodology and application of experimental economics and took charge of field experiments, from which the scholars who attended the class benefited a lot.

In June 2009, China Game Theory Research was registered under the approval of the Ministry of Civil Affairs as China Association of Game Theory and Experimental Economics (CAGTEE). From September 10–11, CAGTEE together with Capital University of Finance and Business, and Beijing Information Science and Technology University hosted an international seminar with the theme of "Facing crises: Frontier research approach in Economics". Professor Egbert Jongen from CPB, Netherlands, Professor D. Houser from George Mason University, Professor Chen Shuheng from

National Chengchi University, Taiwan, and Professor Dong Baomin from Canada made special reports respectively on micro-modeling and its application to welfare reform and employment, progress in experimental economics approaches, agent-based computer economics (ACE) and its application, and advanced game theory. Besides this, scholars in China exchanged their ideas and new research results at the meeting.

Since 2006, the *Game Theory and Experimental Economics Series* has been published annually. Four volumes have come out until now, recording main academic activities, exchange results, and lectures by foreign experts at advanced seminars. The series was organized by CAGTEE, planned by Wang Guocheng, Ge Xinquan and Wang Wenju, and edited by Ge Xinquan and Wang Guocheng.

More than a decade since its establishment has witnessed the growth of CAGTEE. Wang Tongsan, the President of Chinese Association of Quantitative Economics, honorary President Zhang Shouyi and Wu Jiapei, and many other leaders have been consistent in showing their care, interest and support. Relevant government offices and scholars all over China have paid attention to and actively participated in the activities of CAGTEE. For all care and concern we have been receiving, I would like to express my gratitude here. From now on, we will continue to engage ourselves in hosting academic activities about and promoting the application of game theory, experimental economics and related fields.

The Memorabilia of the Input–Output Committee

Liu Qiyun et al.

In 1959, the Chinese Academy of Sciences was the first to study the input–output method. The Operations Laboratory of Institute of Mathematics at the Chinese Academy of Sciences set up an economic team in 1959, which was supported by Qian Xuesen and Hua Luogeng, 2 well-known Chinese mathematicians. The team consisted of 8 members namely Li Bingquan, Chen Xikang, Sun Keding, Zhou Huazhang, Guo Shaoxi, Xu Wenjing, Zhang Chuangu and Yan Shuhai. Professor Hua Luogeng attended the one-week seminar and discussed input–output technology with attendees half a day, every day.

In 1960, an input–output research team was made up by Wu Jiapei, Zhang Shouyi and Gan Zhaoxi from Chinese Academy of Sciences.

From 1964 to 1965, the first enterprise input–output tables were made out by researchers at the Chinese Academy of Sciences. One was 1963 Tianjin Chemical Plant input–output model in physical units, which was completed in 1964. The other was 1964 Anshan Iron & Steel input–output model based on metal mass balance, which was made in 1965.

At the end of 1972, Chen Xikang made a report titled "On Applying Section Aggregate Balance Method to Planning" to the State Development Planning Commission, suggesting the purchase of computers and establishment of a computing center, as well as making an input–output table. SDPC adopted his suggestions, setting up the Computing Center and asking Chen Xikang and others to make the input–output table.

From 1974 to 1976, having cooperated with Beijing College of Economics, Renming University of China and the Computing Center of

State Development Planning Commission (SDPC) for 2 years, Chen Xikang and Yan Shuhai finally succeeded in making the 1973 China input–output model of 61 products in physical units. Their result, the first input–output table in China, was printed and issued interiorly by the Computing Center. Chen Xikang and Xue Xinwei promoted the application of the table from 1976 to 1979 in SDPC. After the reform and opening up, the Institute of Systems Science at Chinese Academy of Sciences made an extended table for 1979 based on the 1973 table. At the same time, the Chinese Academy of Sciences made another 1979 national input–output table according to producer's prices.

At the end of year 1979, the first American economists delegation led by L.R. Klein and K.J. Arrow visited China. Their report contributed 5 pages to explicate the completion of the 1973 Chinese input–output model in physical units and to introduce Chen Xikang. The report, which is now saved in Library of Congress, considered the input–output research carried out by the Operations Laboratory of Institute of Mathematics at Chinese Academy of Sciences as Chinese economic research in the real sense.

In 1979, Chen Xikang from Chinese Academy of Sciences introduced China's 1973 input–output table and its application results at the Seventh International Conference on Input–Output Techniques, which aroused the interest and appreciation from the global input–output academia. His article was later published in *Proceedings of the Seventh International Conference on Input–Output Techniques*.[1] Leontief attended the meeting and made a speech.

From 1980 to 1981, with the help of experts from the State Statistics Bureau, the Institute of Quantitative and Technical Economics at Chinese Academy of Social Sciences and Remin University of China, Shanxi Province succeeded in making a physical units table containing 88 products and a value table of 27 sectors for the year 1979, which became the first input–output table in China. After 1987, most provinces in China could make their own year input–output table following the state table.

---

[1]Chen Xikang, Xue Xinwei, 1984, A non-linear Input–Output Model in Physical Units and its Application in China, *Proceedings of the Seventh International Conference on Input–Output Techniques*, United Nations, pp. 201–209.

From 1982 to 1983, in consideration of suggestions from experts at the Chinese Academy of Sciences, the government decided to make a new national input–output table. The Prediction Center of State Development Planning Commission, State Statistics Bureau and other relevant government offices started to make the 1981 input–output table from the spring of 1982 and completed it at the end of 1983. In addition, the State Statistics Bureau made an extended table for 1983 based on the 1981 table.

In 1986, Professor Leontief, the initiator of input–output theory, paid a visit to China. He had previously worked in China before the founding of the new China. During his visit to China, he showed a great concern for the development of input–output technology in China and spoke highly of what the academia had done to promote the input–output cause in China.

On March 1987, taking into consideration the suggestions from Ma Bin and experts at the Chinese Academy of Sciences, the General Office of the State Council issued *Notification on Carrying out a National Input–Output Investigation* (*No. 3, 1987*), demanding that a national input–output investigation should be carried out every 5 years (year with the end number 2 or 7) and an input–output table should be made based on the investigation. The institutionalization of input–output tables marked a new phrase for input–output study in China. The State Statistics Bureau has made 5 input–output tables for the years 1987, 1992, 1997, 2002 and 2007 since then, as well as 4 extended input–output tables in 1990, 1995, 2000 and 2005.

On March, 1987, the Chinese Input–Output Association, which was subordinate to the Chinese Association of Quantitative Economics, was established through the conjoined efforts of Renmin University of China, the Institute of Systems Science at the Chinese Academy of Sciences, and the National Bureau of Statistics of China.

In October 1988, the First Annual Meeting of Chinese Input–Output Association was held in Jiujiang, Jiangxi Province. Conference papers were collected in *Input–Output Theory and Practice in Contemporary China* was published afterwards by China International Radio Press.

In 1988, a long-term cooperation was formed between the State Information Center and the Institute of Developing Economics in Japan with the aim of making input–output tables among Asian states (regions).

In 1989, Professor W.W. Leontief invited Chen Xikang to write an article about the development of input and output in China. Thus Chen wrote an article called "Input–Output Techniques in China", which was published in the first issue[2] of *Economic Systems Research*, organized by International Input–Output Association.

In 1989, Chen Xikang was the first to put forward the input–occupancy–output technology all over the world. The technology could be used in studying not only relations between the input and output of products of different sectors, but also relations between fixed assets, labor force and natural resources and output of sectors. The technology received great attention internationally. W. Isard, of the Academician of National Academy of Sciences, considered it as a valuable finding and pioneering research. Professor Leontief, the Nobel Prize winner in economics, remarked, "The computing method of input–occupancy–output and complete consumption coefficient was an invention of great significance."

From 1980 to 2008, input–occupancy–output technology was successfully applied to the national grain yield prediction. Chen Xikang through a long time of study put forward a systematic composite factor forecast method to predict grain production. Input–occupancy–output technology was one of the core technologies of this method. Through this method, the average error was 1.9% from 1980 to 2008, which reached an internationally advanced level. Therefore, government leaders spoke highly of the research and conferred on it the First Outstanding Contribution Award in Management Science, the First Outstanding Scientific Achievement Award from the Chinese Academy of Sciences, first prize at the International Operational Research Progress Awards, first prize at the Beijing Science and Technology Awards, and first prize at the Chinese Academy of Sciences Scientific and Technological Advancement Awards.

In July, 1991, the Second National Input–Output Annual Meeting was held in Baotou, Inner Mongolia Autonomous Region. Conference papers were published in *Contemporary Application and Development of Input–Output in China* afterwards by China Statistics Press.

---

[2]Chen Xikang, 1989. Input–Output Techniques in China, *Economic Systems Research*, Vol. 1, No. 1, pp. 87–95.

proceedings named *Input–Output Theory and Practice in China (2007)* were published after the meeting by China Statistics Press.

In 2008, researchers of the Chinese Academy of Science carried out a research and calculation as to the influence of implementing building energy-saving standards on the economy and environment. Their calculation was based on partial input–output closed model and econometric approaches. The research report was spoken highly of by Li Keqiang, the then Vice President of China.

In 2008, State Statistics Bureau and Renmin University of China made fixed-price input–output sequence tables for the years 1987, 1992, 1997, 2002 and 2005, which was supported by Natural Science Foundation of China.

In June 2008, the Chinese Input–Output Association convened the 2008 Input–Output Symposium in Beijing, discussing about the application of fixed-price input–output sequence table from 1987 to 2005. Scholars and graduate students from the State Bureau of Statistics, Renmin University of China, the Academy of Mathematics and Systems Science at the Chinese Academy of Sciences, the School of Management at the University of Chinese Academy of Sciences, Tsinghua University, Tianjin University of Finance and Economics, and Yanshan University attended the symposium.

In 2009, researchers at the Chinese Academy of Sciences carried out a study on industry-classified household consumption potential under the background of the financial crisis. The research put forward countermeasures for different industries to deal with the financial crisis and suggested household consumption should be extended. A report was made accordingly, which was appreciated by Li Keqiang, the then vice president of China. Li hoped that the research results could help boost industry and employment.

In 2009, Tsinghua University, the State Information Center and State Statistics Bureau joined hand in making a regional input–output table based on statistics of 8 regions and carried out research on the dynamic input–output situation in regional economies. Their research got the support from the Natural Science Foundation of China.

In September 2009, the 2007 Chinese Input–Output table was completed. The State Statistics Bureau entrusted Renmin University of China to host the research on its application.

Those who took part in the research included the Academy of Mathematics and Systems Science, the Chinese Academy of Sciences, the University of the Chinese Academy of Sciences, the State Information Center, Tshinghua University, the University of International Business and Economics, Tianjin University of Finance and Economics, the Development Research Center of the State Council, and the Academy of Macroeconomic Research, NDRC.

The Memorabilia of Enterprise Committee (1985–1997)

Hao Chunhe

1985

From October 15–19, 1985, the National Enterprises Symposium on Quantitative Economics and the Foundation Meeting of Enterprise Committee of the Chinese Association of Quantitative Economics was hosted in Anshan, Liaoning Province. Ma Bin, the Counsellor of the Search Center of Economic, Technological and Social Development at the State Council, An Qichun, Researcher at the Research Center of Economics and Management at the State Economic and Trade Commission[1], Li Bingquan, the Deputy Researcher at the Institute of Systems Science at the Chinese Academy of Sciences, and Zhao Xinliang, the Vice Director of the Liaoning Provincial Commission of Economic Planning were employed as Counsellors for the Committee. Yuan Dongzhu, the Chief Economic Manager of Anshan Iron and Steel Group Corporation was elected as the President, and Yan Baozhong, Wu Liansheng and another five people as Vice Presidents. In addition, *Regulations of Enterprise Committee* and the list of secretariat staff were passed at the meeting.

More than 120 conventioneers from 68 companies, scientific research organizations, educational institutions and government offices of 21 provinces, municipalities and autonomous regions were present. More than 60 dissertations were handled in and discussed at the conference.

---

[1] State Economic and Trade Commission was dismissed in 2003 and the Ministry of Commerce was set up in the same year to undertake the same function of the commission.

1986

In September 1986, the Enterprise Committee convened the Second National Enterprise Quantitative Economic Symposium in Kunming, Yunan Province, receiving over 80 people and 60 articles. The conference centered on input–output technology and the research and application of the dynamic input–output model based on the static one. Representatives from Anshan Iron and Steel Corporation Group delivered a report about setting up a metal material balance model by making use of input–output techniques, introducing in detail the steps to make an input–output table. The report was deemed impressive by the conventioneers.

In October 1986, *Enterprise Management and Quantitative Economics* was published by Liaoning People's Publishing House and issued 15,000 copies. This is a book that caters to the modernization of enterprise management in China, as it explicates relations between enterprise management and enterprise quantitative economic research, introduced many kinds of quantitative economic analytical methods such as the input–output method, econometric method, optimal programming method, value engineering technique, and mathematical statistics method, and touched upon the application of computing to enterprise management. Moreover, the book is not only rich in theories but in methods and examples.

1987

In September, 1987, the Enterprise Committee of the Chinese Association of Quantitative Economics convened the 3rd Seminar on the Application of Enterprise Quantitative Economics in Wuxi, Jiangsu Province. This event welcomed 46 attendees and more than 30 articles. Discussions centered on the application of quantitative economic methods such as economic forecast techniques, value engineering, and econometric methods to modern enterprise management.

In October 1987, the Enterprise Committee sent its members to the third annual meeting of the Chinese Association of Quantitative Economics that was held in Wuhan, Hubei Province. Representatives from the committee took part in group discussion on enterprise management, benefit distribution and adjustment in economic system reform, and the application of the input–output method to enterprises. Apart from this, they exchanged reflections on the promotion of the application of quantitative economics. Firstly, the application of quantitative economic methods in enterprises is a practical matter and therefore it requires a high accuracy and reliability. Secondly, apart from a complete and accurate statistics system in the enterprise, the promotion needs communication and cooperation among education institutions, research organizations and worldwide enterprises. Thirdly and more importantly, it demands the enterprise build a research team, only by which quantitative economic methods can take root and develop in enterprises.

In addition, Yuan Dongzhu and Liu Wenquan from the committee were elected the members of the council at the meeting.

1988

In September 1988, the second annual meeting of the Enterprise Committee namely the fourth academic conference was held in Ma'anshan, Anhui Province. The conference elected Yuan Dongzhu as President and made some adjustments to Vice Presidents, members of the council and staff in the secretariat.

The fourth academic conference focused on research on quantitative economic theories, methods and approaches, application and achievements, and problems in application. The discussion was fruitful.

The conference appraised and elected 10 excellent papers from those that were submitted since October 1985, and issued certificates and bonus to prize winners.

1989

In September 1989, the Fifth Enterprise Committee Seminar was convened in Hangzhou, Zhejiang Province. More than 50 conventioneers attended the meeting. More than 20 conference papers were submitted. Discussion and exchanges centered on the application of a comprehensive input index construction production function to enterprise, of a fuzzy evaluation matrix to quality control in materials management, and of econometric methods such as economic benefit forecasts and withdrawing earnings patterns.

1990

In September 1990, the sixth seminar of the Enterprise Committee of the Chinese Association of Quantitative Economics was convened in Jinan, Shandong Province. The Machine Manufacture Factory of Anshan Steel and Iron Corporation gave a report entitled "The Application of Input–Output Model to Metallurgy and Machine Enterprises". At the same time, Anshan Woolen Mill gave a speech titles "On the Input–Output Model of Woolen Enterprises". Moreover, the Fifth Factory of Shanghai Steel Corporation presented two reports, which were "The Effect of Cost–Volume–Profit Analysis on Profit Forecast" and "Application of Linear Programming to Production and Operation".

1991

In June 1991, *Trend in Application of Enterprise Quantitative Economics* of 186,000 words was published by Social Sciences Academic Press. About 3,000 copies were printed. The book consists of 8 parts, namely the application of input–output models, econometrics, cost–volume–profit analysis, linear programming methods, decision-making techniques, optimal programming methods, business operation appraisal, project appraisal, and other quantitative economic methods. Each part is systematic, and rich in content and in examples worth studying.

In September 1991, the seventh symposium of the Enterprise Committee was hosted in Dalian, Liaoning Province, receiving over 70 people and more than 60 articles. People exchanged their ideas through group discussion and speeches.

The third annual meeting of the Enterprise Committee was held before the symposium. The meeting decided that Yuan Dongzhu should be the President of the Committee. It also made adjustments in members of council and vice presidents. In addition, the meeting gave credit to the secretariat of the committee for the work it had done.

1992

In October 1992, the eighth academic symposium of Enterprise Committee was hosted in Xiamen, Fujian Province, receiving more than 60 people and over 60 articles. The conference discussed mainly technical progress and raising labor productivity.

Conventioneers visited Xiamen's economic and technological development zone, gaining knowledge of the situation there through the introduction of those in charge and visiting workshops and seeing their products.

In 1992, Wang Shizhou was awarded the accolade of Excellent Staff by the Chinese Association of Quantitative Economics, and was issued a certificate and a cup.

1993

On September, 1993, the Enterprise Committee of the Chinese Association of Quantitative Economics held its ninth academic seminar in Urumqi, Xinjiang Uygur Autonomous Region, focusing on decision optimization and the application of computer-aided decision-making. More than 60 conventioneers were present and their papers were discussed at the meeting.

Conventioneers visited the Xinjiang Bayi Iron & Steel Corporation and the production line there, by which they were impressed.

1994

In October 1994, the tenth academic symposium of the Enterprise Committee was held in Xinyu, Jiangxi Province. More than 50 conventioneers were present and over 50 articles were submitted. The conference centered on deepening interior reform of enterprises. Zhang Yu, the Secretary of Anshan Iron & Steel committee of the CPC, attended the conference and made a report titled "Refine the Main Part, Diversify Assistance and Deepen Interior Reform".

1995

In October 1995, the 10th anniversary of the founding of the Enterprise Committee and the eleventh academic symposium was held in Benxi, Liaoning Province. The conference focused on ways to get rid of poverty and enliven state-owned enterprise within 3 years. Yuan Dongzhu, the President of Enterprise Committee was present. The Chinese Association of Quantitative Economics sent members to the meeting and delivered a speech.

In October 1995, *Proceedings of Enterprise Quantitative Economics* of about one million and five hundred words was published with 5,000 copies. It was only distributed internally.

1996

In 1996, the secretariat of the Enterprise Committee organized a symposium about the causes of and countermeasures to product cost increases. Some members of the Enterprise Committee and business administrative staff in Anshan Iron and Steel Group Corporation (AISGC) attended the meeting. The conference discussed the cost increases in AISGC products, the reasons for the increases, and measures to lower costs. Relevant articles were published on the 1996–1997 issues of *Journal of Quantitative and Technical Economics* and other magazines.

1997

At the Sixth Annual Meeting of the Chinese Association of Quantitative Economics in July, 1997, Wang Shizhou and Hao Chunhe of the Enterprise Committee were elected as members of the sixth council of CAQE.

Hao Chunhe was issued the 1993–1997 Excellent Staff Award (both certificate and cup) by the Chinese Association of Quantitative Economics.

Part Three

Selected Edit of Speeches at the 30th Anniversary Commemoration of the Founding of the Chinese Association of Quantitative Economics

Han Zhong

Date: May 26, 2009
Location: Jiuhua Spa & Resort, Beijing
Hosts: Li Jinhua, Wang Weiguo, Wang Chengzhang

Wang Weiguo: I am honored to be the host in the period of free discussion. At the 30th anniversary commemoration of the founding of the Chinese Association of Quantitative Economics, we hope all the specialists and scholars have an open discussion and report the achievements in the past 30 years to Mr. Zhang Shouyi, Founder of the association, and Mr. Wang Tongsan, President of the association. In the development of Chinese quantitative economics, the Chinese Academy of Social Sciences has always been a main strength and direct contributor. In addition, colleges and universities are also very important. Let us first invite Dean Zhang from Jilin University to speak.

Zhang Qishan: Today, when we are celebrating the 30th anniversary commemoration of the founding of the Chinese Association of Quantitative Economics, it is of vital significance to link the past and future and to commend founders of the association: Mr. Wu Jiapei and Mr. Shouyi. They are the leading authorities on Chinese quantitative economics, without whose commitment we would never have got this far. Today, we all benefit from them and that is why we show our sincere gratitude to them. The same gratitude also goes to Mr. Feng Wenquan and Mr. Zhou Fang. The Book *Economic Forecasting and Decision-Making* Mr. Wu Jiapei mentioned just now is a compulsory course textbook in university, from which we have benefited all these years. In my opinion, Mr. Zhou Fang is the one who is really engaged in studying mathematical economics for he has written many articles on the discipline, most of which I have read. These two gentlemen have made a great contribution to quantitative economics, which deserves my gratitude. President of the Chinese Association of Quantitative Economics Mr. Wang Tongsan, also devotes himself to the development of Chinese quantitative economics. On the 30th anniversary, I think the best way to commemorate the association and to show respect to the two founders is to carry forward the discipline of quantitative economics. Up to now, in the circle of quantitative economics, some scholars have held the opinion that when theoretical economics in China develops to a certain level, Chinese quantitative economics will have accomplished its mission. I do not agree with this idea for the following two reasons: First, quantitative economics has its reality. Although many disciplines take a mathematical method, none of them can replace ours. For example, econometrics, mathematical economics and game theory are still developing in the US, which indicates the necessity of their existence and reality. We are doing the study of economics from different aspects so we cannot replace each other. Second, quantitative economics is a necessity in China. It has been playing a pioneering role in the development of Chinese economics, especially in the process of catching up with international standards. Currently, most high-level articles refer to quantitative analysis. Western economics embraces thoughts of various schools, the introduction and extensive study of which can be attributed to quantitative economics. This is because we, the quantitative economists, keep studying, introducing, adsorbing and adapting, on the basis of which we improve and develop, so that economic disciplines blossom in China and reach the international standards. Quantitative economics is of critical importance not only in its own development but also in fields of Chinese economic disciplines and management disciplines. In this way, as quantitative economists, we should cherish what we have achieved in the past and enhance academic exchanges. It is our incumbent duty to give full play to quantitative economics in practice and we are confident to fulfill the mission.

Today, we gather merrily and take a historical review to summarize experiences, which will give an impetus to the development of Chinese quantitative economics. As a quantitative economist, I hope all my peers (including myself) should make efforts in two aspects: One is academic research. Only as our academic level achieves status at home and abroad can we better cement the place of quantitative economics in China. The realization of this grand objective requires academic research. I hope, in the future, a Nobel economics laureate in China will be a quantitative economist. As for the different development between other countries and China, it can be seen as a contribution to the World to sum up the experience of Chinese economic development. The other aspect is that we, the present can hold administrative posts to the greatest possibility. Only in this way, can we have a say in government to better consolidate the status of quantitative economics, making realistic contributions to Chinese and world economic development as well as inheriting the career pioneered by the predecessors.

Feng Wenquan: First of all, I extend my warm congratulations to the 30th anniversary of the founding of the Chinese Association of Quantitative Economics. During the past 30 years, our quantitative economy has boomed at an amazing speed. If we compare the Association of Quantitative Economics in 1984 to now, we can see the differences. In 1984, we introduced the quantitative economy just as a new concept and new method with a limited talent team. But today, courses in quantitative economy have been offered in so many colleges and universities. At a rough estimate, there are about 100 colleges and universities which have course on economic projection. In schools at a national scale, courses on the quantitative economy occupy a key position. The quantitative economy can be found everywhere in business circle especially in finance. We used to be unfamiliar with stock, but thanks to quantitative economics, we are familiar now. In terms of the mission of the association, we admit that quantitative economics contributes a lot to the national development; however, our mission is still yet to be accomplished because to satisfy the need of rapid national development, there are too many problems to solve such as decision-making. Nowadays, there are more big companies as well as more small ones, which require an accurate and quick decision-making process based on time, such as solving the financial crisis. In 2008, the American subprime mortgage crisis broke out, causing heavy losses to China, especially to the Bank of China (BOC) and to China Construction Bank (CCB). The total losses are not clear yet. Statistics from America show we suffered badly but ours show different. After the situation, China Merchants Bank (CMB) made a flexible decision while BOC and CCB were slow in decision-making, causing heavy losses. Decision-making plays a very important role to senior managers. All these problems need to be fixed, but in which we have little research. As for the textbooks I wrote, there are many problems. For example, much needs to be modified according to recent situation in modern society. The accuracy and speed of decision-making is contradictory as the depth of decision-making is related to its level and scale. China is not what it was 30 years ago. We are now taking in an active part in the world decision-making process. After the financial crisis, Chinese influence will be enhanced. Our mission is far from accomplishment so our Chinese Association of Quantitative Economics should make new development planning after 30 years' growth, and should have a new starting point to accelerate our development. We should speed up the development of quantitative economics. It has been increasingly difficult to make economic projections, especially the specific and detailed problems big companies will be faced with. As an association, we need to have our own development plan in the new situation. There is a clear gap between our growth rate and modern requirements so there is a long way to go. The day when quantitative economics becomes a mainstream discipline is sure to come.

Li Xuesong: Chinese quantitative economics has achieved great success by achieving its 30th anniversary and has drawn the world's attention for the spectacular extension and depth of research. In the next 20–30 years, the Chinese economy will undergo a big change: There is an increasing possibility that China will exceed America in economic aggregate, with a more rapid growth in economic strength. Against such backdrop, quantitative economics will have a bigger stage. Internationally, more scholars and overseas students from China who have accomplishments in quantitative economics will come back to China to promote its development. Our generation suffers great pressure in spite of our experience abroad and promising opportunities. We should conform with the quantitative economics to the historical trend of the time, unite all its strengths and introduce its talents and methods. In the US, econometrics enjoys a prominent position as one of the three main subjects for doctoral degree study in top-ranking universities. Up to now, econometrics includes more and more extensive content, from classic methodology to time series and microcosmic measurement, all aspects developing at a fast speed. Quantitative economics in China contains not only American quantitative economics but also methods of mathematical economics, input and output, and large-scale model. The achievement of reform and opening-up is largely due to the impetus from economic circles. The function of economics is to put forward market trends, defend the reform of marketization with determination, defend socialist market economy and firmly promote globalization. In the following 20–30 years, the economics should be specified further with more deep and specific research. The well-known senior quantitative economists have put forward the objective of market economy reform which led to fruitful results. In the coming 30 years, we need to do more specified research with leading methods from quantitative economics. Now the Chinese Academy of Social Sciences often receive tasks from the Party Central Committee and the State Council, which ask us to provide quick countermeasure study and simulation, showing the great importance and high demand of central leaders attach to our discipline. However, we often feel the tasks beyond our reach. The methods are not updated at the speed needed and some problems are hard to handle. The national development generates enormous need and immense potential for us and so does the global development. Now the Americans are writing a series of quantitative economics manuals, each of which contains the research in recent years of different scholars in quantitative economics. These manuals demonstrate the development history of quantitative economics as well as the disparity between them and us. As a result, we, all the members in the association, scientific research institutions, colleges and universities, should work hard to give full play our fine traditions, unite as one, making concerted effort to propel the development of quantitative economics.

Wang Naijing: I am very glad to take part in the 30th anniversary commemoration of the founding of the Chinese Association of Quantitative Economics. The discipline of quantitative economics and the reform and opening up are synchronous. At its early stage, it was all about learning, introducing and promoting. Today, at this time, we should switch our focus to its future. There are many self-established disciplines in China which are in decline but our quantitative discipline is becoming increasingly vigorous with ever-improved annual conferences. I summarize the reasons for this in 6 points: (1) Our predecessors in quantitative economics have laid a solid foundation for us with their capabilities and focus on future development, which makes a fine basis of working. (2) We have a capable President and Secretary General. As an association, these two positions are of the most importance. All the previous presidents acted as champions of quantitative economics banners and secretaries general were more than competent. It is very important for them to be banners of cohesion and influence. (3) The association has a core academic team. We gather in the association, boasting core specialists and professors, where everyone loves the subject. Academic regular conferences are the foundation of our association. We should make good use of administrative resources, which will help develop our subject. (4) We have a host of teachers and students who love quantitative economics. They inherit the subject and do research on it. (5) The subject itself is energetic. Our subject started at the same time as the reform and opening-up policy. The reform has not succeeded yet so our subject has not reached its peak, which leaves space for further development. The annual conferences have witnessed growing numbers of participants every year, which exhibits the energy it has. Our mission cannot reach an end and it has not ceased despite the attacks. (6) We have a good journal, *Quantitative & Technical Economics*, which is a national and top-class journal of extensive influence among colleges and universities. These six reasons are the fundamental basis for believing our subject will continue to flourish. I am not saying our subject is perfect without any problems. Instead, we need to have a discussion. I think there is a problem in the training of students. We have not laid a solid academic foundation for students from the Department of Economics. We have not reached international standards in basic education in teaching quantitative economics. We only cultivate students to an undergraduate level, so when students study for a doctor's degree, they may feel it beyond their reach. The basic education of quantitative economics is limited and the training has problems. The second problem is the quantitative economics is disconnected from industry. We used to make input and output for industries and we used to promote optimization in industries, but now we seem far from them. Industry no longer attaches importance to the methods of quantitative economics, and our subject penetrates less and less in industry. The third problem is that we value application over theory.

I think, on the 30th anniversary commemoration of the founding of the Chinese Association of Quantitative Economics, we need to further develop the Five-Year Plan and Ten-Year Plan. As long as the reform and opening-up has not stopped, neither will the quantitative economics. Thanks to all.

Liu Qiyun: First of all, I would like to show my sincere respect to the founders of our association, Professor Wu Jiapei and Professor Zhang Shouyi, on the occasion of the 30th anniversary of the founding of the Chinese Association of Quantitative Economics. Thanks to them for setting up this platform designed for professionals studying and teaching quantitative economics. After 30 years of effort, the association has now developed into a large team. The research questions and team scale in the past cannot compete with those of today, for which I would like to show my respect again to the predecessors who have been leading us. My major field is in input and output research. As a professional committee, the Input–Output Committee has done a lot of work. Professor Zhang Shouyi and Professor Wu Jiapei mentioned the work in the review. The development of input and output is proceeding with the purpose of serving the national economy. The current development is still facing great resistance, just as Professor Wang Naijing mentioned it has been hard to excite industry. Now we need to find a new point of penetration. The current situation is posing a big challenge to quantitative economics. What we are facing today is very different from the past. Especially in the recent year, the financial crisis has posed a challenge to us which requires the mentality and methods of quantitative economics. The situation propels us to work harder and grow faster. In our circle, creativity is the only way out. If the problem is different, the measures should also change. We should not only base our achievement on the past but also cultivate the young generation. Now there is an impetuous mood in the academic circle. It is difficult to think and work in a down-to-earth manner. I hope young people can sit at the table, resist the temptation, and learn from senior scholars in thinking, creativity and the pioneering spirit. What is more, the Input–Output Committee holds conferences every three years, and the numbers of the input and output research team is reducing, but we are confident of doing a good job. We have always had connections with international colleagues and we strive to push the discipline of Chinese input–output to a higher level. Finally, I would like to show my respect again to all the senior scholars and specialists, thank you!

Wang Guocheng: Thanks to the host. Today is an exciting day and a memorable day for quantitative economics. I would like to show my high esteem for the founders of our association, Professor Wu Jiapei and Professor Zhang Shouyi. As for me, there is another reason, Professor Zhang Shouyi is my direct doctoral supervisor and the love and kindness my supervisor gives to me makes me owe him a great debt of gratitude. I am going to talk about the development of game theory. With support and guide from senior scholars, our initial move was enlightened by Professor Zhang Shouyi's proposal. In 1996, the First Symposium of Economic Game Theory was held in Hangzhou. In 1998, with the approval of the General Council, a Professional Committee of Economic Game Theory was established, named the National Association of Game Theory. As the needs of development increased, especially those of experimental economics, the committee was renamed as the National Association of Game Theory and Experimental Economics. So we are paying more attention to the development of quantitative economics and we have had several academic conferences about the issue in the past few years. In addition, I would like to share my reflections on the development of game theory and quantitative economics. Over the past 30 years, quantitative economics has been penetrating into and even occupied forward position of national mainstream economics, which would not realized without our endeavor. Yet, there comes another problem which lies in the breakthrough points of the sustainable development of the quantitative economics. Our advantages have been diluted and that is why we need to focus on these to find out the concentrated breakthrough points for further development. That is all. Thank you!

Li Linshu: My respectful scholars, on behalf of Central Radio and Television University (CRTVU), and radio and television universities (RTVU) in the country, I would like to express congratulations to the Chinese Association of Quantitative Economics for its 30th anniversary. Special thanks also go to Mr. Wu Jiapei and Mr. Zhang Shouyi as lecturers and chief editors at CRTVU. They took part in the discipline construction and professional building 10 years ago. Thanks also go to the professional committee for their consistent support and help. I signed up for this conference a long time ago and have been looking forward to its opening. This year is also the 30th anniversary for Radio and Television University (RTVU). After Deng Xiaoping instructed that preparations be made for the establishment of RTVU in February, 1978, the university opened on February 6, 1979. This matched with the reform and opening-up policy as well as the founding of the Association of Quantitative Economics. Dramatic changes have taken place in RTVU in the past 30 years. I would like to take this opportunity to report briefly what it is like now. RTVU contains 1 CRTVU, 4 provisional ones, about 1,000 ones of prefecture level, and 2,000 ones of county level, covering all areas of China, Tibet included. The number of teaching and administrative staff is 120,000, of whom 80,000 are teachers. The large team undertakes the task of mass education, involving team construction and discipline construction. This is the biggest school in the world with over 3.2 million students at the school (continuing education, including education with or without record of formal schooling). The current situation of the school development is owed largely to the Chinese Association of Quantitative Economics, who participated in the general committee as a collective organization CRTVUS being the main body 10 years ago (in 1998), and established a professional committee CRTVUS being the main body. Thanks to the association for such an associate idea and attention to mass education.

Here, I sum up several special contributions the association has made to supporting TV University and educational reform: (1) It strengthens the awareness of academic research for mass education at TV University, especially deepening the understanding of the important role quantitative economics plays in theory and related courses. It makes sense for receiving bodies: Over three million students (one million students majoring in economics and management, taking up 1/3) with over 100 majors. In the reform of discipline construction and professional building, an understanding of quantitative economics plays a key role and also nurtures an academic atmosphere of valuing quantitative analysis. (2) It sets up a platform for academic research. RTVU is relatively weak in academic research. Fortunately, with support from the general committee, the platform, which was established after the professional committee was set up, has made it possible for RTVU to develop quantitative economics as well as other disciplines in the past 10 years. On this platform, related academic research and research projects are doing well. RTVU gets lot of projects directly from the Ministry of Education, including one about benefit–cost analysis of distance education which won the gold award in the international distance education circle. The project is conducted under the guidance of the Chinese Association of Quantitative Economics, especially of Professor Li Yiyuan. (3) The General Committee has propelled the educational reform in RTVU, especially in discipline construction and curricula construction. The Chinese Association of Quantitative Economics strengthened courses of quantitative economics in the majors of economics and management in 1996, thanks to Professors Li Yiyuan and Li Zinai who were invited to teach an introduction of the knowledge economy as part of the economic mathematics course; and Professors Zhang Shouyi and Wu Jiapei who gave lectures themselves as part of the course. For example, RTVU is now actually based on distance education or modern information technology, especially using internet education, which moves RTVU onto a modern track based on multimedia especially comprehensive application of computer networks. This transformation did not start until 1999. In campus conferences in 1998, especially in the face of economic mathematics broadcast by CCTV or CETV, Mr. Wu Jiapei proposed that information technology (IT) will be a new thing worth promoting in curriculum reform, especially distance education based on IT. He mentioned IT in the beginning of 1998. After that, modern distance education was specifically proposed in 1999 by the country, which proved Mr. Wu's foresight. Mr. Zhang Shouyi specially compiled the textbook *Introduction to Knowledge Economics*, for us. RTVU promoted the course early as a popular course when Jiang Zemin mentioned the knowledge economy. The concept of knowledge was not just a course; the contents of it accelerated the pace of curriculum reform. With support from all the specialists, the system of RTVU especially CRTVU has made progress in scientific research, such as 8–9 national excellent courses every year. This is not easy for RTVU, so special thanks go to Professors Li Zinai and Li Yiyuan. Finally, the support from General Committee has actually laid solid foundation for mass education in RTVU, especially its future educational reform. The State Council and Ministry of Education are now making medium and long-term program outlines for educational reform, which has given three historical tasks to RTVU. The first is to establish open universities covering all areas in China based on RTVU, which will provide a platform for continuing education and life-long education. The second is to establish a platform of public service for national continuing education relying on RTVU. Universal education after formal education will come into play on the platform. The last one is to build a national center for digitalized teaching to serve the whole country. With support from the General Committee and fruits of the previous reforms, we will lay a fine foundation. Finally, we have some hopes and suggestions. First, we hope the General Committee will help us register the name of the professional committee as soon as possible. Second, we hope all the specialists can pay attention to mass education while still focusing on high academics. Academic achievements, academic concepts and ideas rely on master's to be put into practice, which will exert an immeasurable potential influence upon the three million students, people around them and quantitative economics, as well as its role in boosting national economy. We hope all teachers can pay attention to us and transform your achievements in scientific research. Third, we hope the General Committee can offer opportunities to exchange in all aspects including substantive cooperation and link to cooperate, benefiting us all in a positive way. Thanks again to General Committee and all the specialists. We hope the Chinese Association of Quantitative Economics can become better with each passing day.

Wang Wenju: I am very fortunate to participate in the conference today with all the present senior scholars. First, I would like to show gratitude to our association, without which Capital University of Economics and Business (CUEB) would not be what it is today. Quantitative economics started early in CUEB. In 1978, Professor Xue Ying was one of the 18 teachers. Besides Professor Xue Ying, we have Professor Yuan Fengqi, Professor Qi Xianglan and Professor Lin Yin, leading authorities of Chinese Association of Quantitative Economics, who have made great contributions to our school. Quantitative economics is 30 years old in CUEB, and its doctoral program is 10 years old. Thanks to quantitative economics, the quantitative economics doctoral program and another 10 doctoral programs are conducted. Without quantitative economics, CUEB would not be what it is today. I would like to take this chance to show gratitude to everyone and the national circle of quantitative economics for their support and help given to CUEB. Second, the Association of Mathematical Economics was originally directed by Professor Yuan Fengqi and now I am in charge, offering services under the guidance of Professor Zhou Fang. Professor Wang Guocheng, Professor Ge Xinquan and I have held some conferences by gathering members of the Association of Mathematical Economics and Association of Game Theory, which had positive effects. We hope to make the association better in the next phase. Third, business management including management and economics has been using quantitative economic methods recently. So it is suggested the Association of Quantitative Economics strengthen research on theory and methodology, based on quantitative economics to promote economic development. In addition, my suggestion is to build bases of association assigned by disciplines according to the different qualities of different schools. Every year, a small symposium of 30–50 people is held, focusing on one discipline to promote different subjects in China to catch up with international standards, which will be good for the Association of Quantitative Economics. What's more, I hope our CUEB can better serve all and offer more help. Thanks to all!

Wang Chengzhang: Now let us open the floor for discussion this afternoon. First, I suggest that Professor He Luzhi from Xinjiang University speak for he is one of the members who attended our association at the earliest stage (in 1982).

He Lunzhi: I am a veteran of quantitative economics but cannot compare with Professor Zhang in seniority. It is a good time when the Chinese Association of Quantitative Economics celebrates its 30th birthday. My academic specialties, if any, cannot be separated from the support and guidance of the association. Here, I show my greatest gratitude to all senior scholars. Next, I am going to talk about three tips from my understanding:

First, quantitative economics is now enjoying popular support. I remember someone said in the conference held by Xi'an PLA Institute of Politics that it's our hope that after 30 years of construction, Chinese quantitative economics will reach world level. Now that 30 years have passed, there are three indicators to examine as to whether we have reached that goal or not. First, in terms of discipline construction, nearly all schools with majors in economics or management have opened related courses in quantitative economics whether they are comprehensive universities or engineering colleges or other normal universities. So at this point, discipline construction can be regarded as successful. As for the second indicator, I was glad to have attended the conference on the national political economy about Marxist methodology, which was held by Professor. He Linsheng from Northwest University. A teacher responsible for political work said theoretical research can we have innovation. Of course, empirical research can have innovation, and it should have, but it must be combined with theoretical research.

I think the following two points will be very significant to the sustainable development of quantitative economics. Up to now, while China is catching up with the world, the basic education of quantitative economics should be internationalized. Now some students of ours show interest in theoretical research but sometimes I feel they are not trained enough and they have not had enough training on theoretical research. Foreign doctors of economics spend at least 5 years, even 6 years. They spend first 3 years taking courses while domestic doctors spend just 1 year in courses and then move on to do training of theoretical research, which is obviously insufficient. How to solve these problems? How to train doctors especially those of quantitative economics? I think there are many problems to handle.

Zhu Pingfang: I felt obliged to come to this conference after I got my invitation, firstly because I have not seen Professor Zhang, who has offered us so much help, for years. I first attended the annual conference of the Chinese Association of Quantitative Economics in 1990. The location was in the guest house of the Ministry of Chemical Industry. I have attended nearly every conference thereafter. In the conference, Professor Zhang showed great care for me and the Shanghai Association of Quantitative Economics, as well as Shanghai University of Finance and Economics (SHUFE). So this time, on the one hand I stand for the Shanghai Academy of Social Sciences (SHASS); on the other hand, I extend greetings to Professor Zhang, on behalf of the Shanghai Association of Quantitative Economics as well as professors from SHUFE, Fudan University, East China Normal University (ECNU) and Tongji University who are familiar with Professor Zhang. Every time I attend the annual conference of the Chinese Association of Quantitative Economics, I make a lot of friends and get support and care from the chairman of the association. Professor Zhang often gives suggestions for our discipline construction. I remember Professor Zhang came to Shanghai for some investigation and survey work in 1995 at the Shanghai Information Center (SIC). Wu Weiyang from Department of Forecast in SIC was responsible for the reception. He also sends his regards to you. Professor Zhang did another survey in SHASS and was received by Li Wen, who is now Vice Chairman of the Chinese People's Political Consultative Conference (CPPCC), and one of the earliest professors in Shanghai to study quantitative economic. Activities related to quantitative economics are continually held in Shanghai. Recently, the Shanghai Association of Quantitative Economics held a conference on changing the term of office, the sixth of its kind, in which we re-registered nearly 200 members from colleges and research institutions.

I heard talk on the issue of sustainable development of quantitative economics this morning. As for the question of could quantitative economics disappear if theoretical economics develops, I do not think it's possible because quantitative economics includes many quantitative methods in terms of its development field. Everyone focuses on different things so it is impossible to replace econometrics and quantitative economics. I helped one of my supervisors to review a student's paper as the paper contained a lot of quantitative stuff. I asked the student: "Why do you put them in your paper?" He said: "It won't work without them". Now here comes the question of how to use quantitative economics? After I saw the paper with quantitative content added, I found problems. To be positive you could call this progress, but to be negative, it could be termed abuse of the subject. But at least I knew one thing: He knew the importance of quantitative economics. In other words, economic research cannot work without empirical research and quantitative instrument to support the conjecture, and that's the significance of quantitative economics. We can see the development of association from my school: SHUFE used to enroll 2 or 3 students for the quantitative economics master's degree every year; we enrolled 3 students in 1999, 4 in 2000, 12 in 2001, 26 in 2002, many of whom majored in economics, engineering, computer and mathematics with first application of quantitative economics.

I am very glad to be at this conference to see Professor Zhang who is devoted to the construction of quantitative economics, and he is in good health. We are also very thankful to Professor Ge Xinquan who came here particularly for the conference. We show sincere gratitude to the association, Professor Ge Xinquan and Beijing Information Science and Technology University for making it possible to gather us together to celebrate the 30th anniversary of the founding as well as the future development of the discipline.

Chen Nianhong: All specialists and representatives, today we commemorate the 30th anniversary of the founding of the Chinese Association
of Quantitative Economics here, where I see many admired specialists.
I took a lecture in Anhui University in 1984 when Professor Li Bingquan
promoted input and output in the country. Since it is the 30th anniversary,
I would like to share my personal feelings about the development of the
association. The annual conferences have been better year after year under
the direct organization, direction and facilitation of the secretariat, forming
a platform for exchanging and interacting and with information for masters,
doctors and young scholars. The discipline has a history of 30 years with
a talented team. Here I would like to propose that our association should
be better known to the society or in other words, we should make our
association more energetic by making use of our talent pool, information
base and data base to play a more important role in a new way. So I suggest
that like the "Bluebook" published by the Chinese Academy of Social
Sciences, our association should be more flexible with joint efforts for
it is a non-governmental organization. At the same time, we should have
authoritative information especially on the operation and development of
the social economy. I ran into an administrative officer who said "Since
you study economics, can you tell me when the turning point of economic
recovery will come? In which month? Even though you are wrong or give
an absurd answer without any scientific basis, you have to give me an
answer!" This example tells us all levels, stratum and even ordinary people
need economics, especially quantitative economics to be voiced and to be
published. This made me think: Professor Xie Fuzhan used to speak often
when he was in the National Bureau of Statistics but since he became
director of the Development Research Center of the State Council, not
anymore. Even though, he does not make speeches anymore, we all want to
hear him speak. So I suggest our annual conference should not only publish
collected papers, but also the association should become a singular voice
with social responsibility, just like the American Economic Association.
We have so many specialists, scholars and professors so we should organize
a special subject at the annual conference to voice opinions on the medium-
and long-term development of the Chinese economy promulgated by media
release. Our Association of Quantitative Economics will organize and guide
articles of free discussion into systematic and specific data, which will

inject vitality into our association and make it more widely known to society.

Gao Tiemei: Our Dean Wang Weiguo from Northeast University of Finance (NUF) left by air. Before he left, he authorized me to give a speech on behalf of NUF. First of all, I feel lucky to attend the 30th anniversary Commemoration of the Founding of Chinese Association of Quantitative Economics. On behalf of Dean Wang and professors and students of quantitative economics from NUF, I would like to show our high esteem to the Chinese Association of Quantitative Economics for its 30th birthday as well as to senior quantitative economists of the association! The past 30 years have witnessed the development of quantitative economics, which can be attributed to the commitment of quantitative economists. We hope the association will thrive and prosper. We have now trained more and more talented people and possess a bigger team, all of which comes from the endeavor of senior quantitative economists in the association. What's more, our junior scholars grow up with the association and so does the younger generation.

I would like to talk about the development history of quantitative economics in NUF. In 1984, NUF established the Institute of Quantitative Economics and the quantitative economics master's program in 1985, which was among the first in China. We started our doctoral program in 1998 together with other 5 universities: Renmin University, Shanghai University of Finance and Economics, Capital University of Economics after the first batch of Jilin University, Tsinghua University and Chinese Academy of Social Sciences. In 2001, our quantitative economics became a key discipline of Liaoning Province and a key research area of Liaoning Province in 2007. We have set up a research base for the analysis and forecast of economics and econometrics as a Chinese Academy of Sciences branch center in the northeast. Thanks to the establishment of the doctoral program, over 40 teachers in our department, including other teachers of quantitative economics, will achieve a doctoral degree, accounting for about 40%. Because more and more young teachers get their Ph.D., and a large number of students will graduate in a few years, teachers with doctoral degrees will account for about 60% to 80%. Now we have 9 doctoral supervisors. Our course in econometrics was rated as a National Excellent Course two years ago, as well as key discipline for cultivation. Up to now, the major of quantitative economics in NUF is still flourishing, but with a low threshold comparing with other key universities like Sun Yat-Sen University, Xiamen University, Jilin University, Tsinghua University and Huazhong University of Science and Technology. We have to catch up with them to narrow the gap.

I am going to talk about my own feelings. My growth is closely connected with the Chinese Association of Quantitative Economics in the 30 years. Senior quantitative economists care much for our growth. Professor Zhang Shouyi always has time for me. My first national youth project on social science was reviewed by Professor Zhang. Though the project had limited funding, it helped me grow a solid foundation. Every time we ask Professor Zhang questions, he is always approachable and guides us patiently and systematically. At the same time, Professor Wu Jiapei also approves of our work and points out the direction for further development. Despite his busy workload, he is always extremely earnest and patient when young students ask him questions. We should learn from the senior scholars and treat our students in the same earnest manner. We got much help and guidance from our peers in research process. In 1998, Professor Dong Wenquan's sudden death was a big blow to us because he was our academic leader. At that time, the Ministry of Finance asked us to make a large-scale simultaneous equations model to simulate fiscal policy. Due to our limited experience, we visited 7–8 organizations in Beijing. When we met Wang Tongsan, Director of the Chinese Academy of Social Sciences, he was so warm-hearted and convened all researchers of modeling to help our analysis. At that time, we all relied on our self-made programs, but he told us to use ready-made software to achieve double results with half the work. Under his guidance, we spent much less time to make the models, switching our focus from studying programs to studying models, which saved much time for us. What's more, Professor Shen Lisheng taught us what he just researched without reservation. Later on when I asked questions to Professor Liang Youcai in State Information Center, I was moved as he reviewed the equations one by one like correcting homework and told me about the problems of the coefficient and which variable was missing in the equation. The annual conference of the Association of Quantitative Economics provides us with a platform where we can take our master's students and doctoral students to learn stuff. Besides theoretical development such as the high quality published papers, we also give advice to government, which was the goal when Professor Dong Wenquan led the team. We succeeded in predicting the "valley bottom" of 1989, which was reported to leaders of governments at all levels by Professor Dong, however not taken seriously. At that time, the overheated economy was caused by the civil mentality of inflation while the economy itself was not overheated so austerity measures should have not been implemented but were, which led to the negative economic growth at the end of 1989. The National Bureau of Statistics asked us to calculate and the result was said to be taken to the Central Working Conference. Li Peng, then-Prime Minister told Ma Bin to write a thank-you note to Professor Dong. The current conferences in spring and autumn can hear our predicted outcomes there, which take good effect and should be added weight as we can provide a reference frame for the country's decision-making.

Xu Xiaoguang: Today is a memorable day and I am glad to attend this meeting. First of all, I would like to show my high esteem to the founders of quantitative economics. My sincere gratitude goes to Professor Zhang Shouyi as I was personally guided by him and he was the chairman of the oral defense committee for my doctoral dissertation. All present here are senior scholars in quantitative economics, who have made outstanding contributions to the subject, so I would like to show my respect to you for bringing me to the research field. I am from Shenzhen University, where this annual conference is held. I am sorry for the hasty preparation, limited experience and poor treatment. Shenzhen University was not built until 1983 and I came here in 1999. Before that, there was no course or teacher for quantitative economics in our school of economics. I began to teach econometrics in 1999, both for undergraduates and graduates. For the period between 1999 and 2006, I was the only teacher of econometrics and this was a heavy task. In the lecturing, I found students were weak in basic theory because they only took courses of higher mathematics and linear algebra plus a little mathematical statistics, which made econometrics a challenging task; a problem facing all universities and colleges. In 2007, I pointed the problem to the campus dean's office and suggested we set up a dual-degree class of mathematic finance, enrolling mainly students of finance and mathematics by exams. The class I am in charge of has been opened for 2 years with main courses including mathematical analysis, higher algebra, probability theory, mathematical statistics and

We have a master's program of quantitative economics in our school, but it started late. As we are in the School of Economics and Management, we have to find a link to bridge economics and management, on which we have achieved much in the recent 2 years. Vice President Han Qiushi said I got the National Prize for Progress in Science and Technology in 2008 because there was a theory of knowledge management and a model of econometrics in our project, Platform of the Evaluation Technique for Toxic and Harmful Substances, which took effect. He said we won the National Prize for Progress in Science and Technology again on May 22 and another piece of good news this afternoon is that our School of Economics and Management has also been submitted for the prize.

On the whole, we have been doing well in the 2 years with obvious effects resulting from the platform quantitative economics has built for us and the knowledge management I have been doing these years. In addition, it must be remembered that our university is a newly built one. The former Beijing Institute of Machinery was the smallest school in Ministry of Machine Building, whose institution status was changed under the charge of Beijing Municipal Education Commission and then merged in 1998. In general our university is a new one. I would like to extend gratitude to all the specialists for their care and help during our growth. I hope to get further support from you in the future if possible and invite you to come to our university to guide our work.

Next, I am going to talk about my opinion on quantitative economics. As all the specialists have just mentioned, the Chinese Association of Quantitative Economics was initiated by Senior Quantitative Economists like Professors Wu Jiapei and Zhang Shouyi, which has been of distinct Chinese characteristics like introducing, digesting, adsorbing and innovating from the start. The Chinese Association of Quantitative Economics has played a role of increasing importance after 30 years of development, especially in responding to environmental change such as the financial crisis. In the recognition and reappraisal of economic theories, economic methods and classical models, and on innovation, quantitative economics cannot be ignored.

The first problem I am going to talk about is the misuse of the application of models, which is a problem universities and colleges are going to tackle because this is a systematic educational system. When we teach systematic knowledge like econometrics, economics including mathematical economics and game theory, we should stick to the truth. The school is partly responsible for the misuse of models. When I went to a Ph.D. defense in a university, I turned randomly to one page on which there was a regression equation with stochastic terms. I asked:" How can a regression equation have stochastic terms?" He answered:" I copied it." The abuse does exist and we should be responsible for it when teaching quantitative economics. I think we can get over the problem gradually by being strict with ourselves in teaching and research. It is normal to have problems in the process of innovation but basic concepts, basic terminology and basic connotation should not be misused. Another problem was also mentioned by specialists in the morning. Many people were not able to use the method of quantitative economics 30 years ago due to limitations but there comes the problem of what advantages do we quantitative economists have when 30 years later many people use mathematic methods in research? And how brilliantly we use methods of quantitative economics? I think our advantage lies in the innovation of quantitative economics, specifically as follows: It is economic theories that matter as what we are studying are economic problems; besides, we need to probe deeply into realistic social surveys. Sometimes when we do research on a certain issue, we need to use some data but existing data do have problems. It is very important to do field research of data. What's more, in the research we should focus on the innovation of models and techniques, especially in creating new models and tools. Tool and theory are in a dialectical relation. It is key to have ideas to study economic problems, without which even the most mathematic instrument cannot work. Our strength lies in making a precise location of the idea by understanding the problem from its past, current situation and future and then making it instrumentalized and mathematized.

China has put forward the strategy of independent innovation. There are three aspects of the strategy that are suitable for the situation; one of them is original innovation. Just as previous specialists mentioned, due to a promising domestic environment after the reform and opening-up, many returnees came home with a solid foundation in modeling and good understanding of western theories, which made it possible for them to accomplish much in the model innovation of economic quantities. This is what original innovation is. Another one is integrated innovation.

Now there are many model tools and Professor Zhang summarize them into three categories this morning, with many more sub-classes in each category, so it is key to have integrated innovation, which requires us to integrate all models according to realistic problems as each model has its own advantages and disadvantages. Another reason is that the whole of economics deals with the problem of resource combination and the application model has the same problem, so our innovation can take this route. The third way is to introduce, digest, absorb and innovate. We do need to learn from foreign countries especially in measurement models. In spite of the progress and achievement, there is still a wide gap between foreign countries and us. Before we learn well the old knowledge, they have new innovations. So I think it is important to digest, absorb and innovate. The innovation problem does exist in transplanting western, in pure market economy tools into a market economy with Chinese characteristics. This is the innovation problem I am talking about.

The second problem is the several categories of the models Professor Zhang mentioned in the morning. Personally speaking, we need to focus on simulation analysis, which has a lot of contents such as system simulation and experimental economics. Several national ministries and commissions have collected mass data but how to do data mining? The first thing comes to my mind is simulation. CASS is now doing such work, and getting substantial material support from the Ministry of Science and Technology, which demonstrates the government's focus on how to do data mining. It is insufficient to rely on present measurement models, so the third model Professor Zhang just mentioned will play a vital role, which involves comprehensive problems such as computer software, natural science, and economics. From my aspect of knowledge management, measurement models and mathematical models are tools of knowledge mining, the combination of which is good for mass data mining. Of course, traditional statistics is a tool for data mining. I got enlightened by Professor Zhang this morning and I think we need to focus on three major parts in our next step of work. The first and second parts were relatively well developed while the third part requires us to do much work, especially joint work. Due to the heavy workload, we need to join together, and the Chinese Association of Quantitative Economics can play its role to integrate national power. The government has invested much into the project, with every ministry or commission investing 70 million to 80 million on each platform. Before long, we built the platform for the Ministry of Human Resources, and the Social Security and Ministry of Finance gave us 80 million yuan for hardware and software. We people who study quantitative economics will play a vital role in the next step because there are many submerged problems like employment and social security in constructing a harmonious society, dealing with which requires mass data. What we need to do is to use advanced analytical tools to do mining of value. That is all. Thank you!

Zhang Shouyi: I benefited a lot from this 1-day symposium. The financial crisis was mentioned, which is just another example world economic crises since that of 1929–1933. The inescapability of economic crises demands us to strengthen research on it. Keynesianism was born after the Great Depression of 1929–1933 with the publishing of *The General Theory of Employment, Interest and Money* in 1936, which produced macro-economics, moving economic thought to a new level. Macroeconomic regulation and control in all countries cannot be done without Keynesianism. Adam Smith cannot rely just on "one hand" and neoliberalism has become a nuisance hated by everyone. I'd like to call everyone's attention to the question of what changes will take place in western economics? Is it possible that there will be new "Keynes" to lift economics to a new level? So please focus closely on the newest trends of mainstream publications both at home and abroad. I suppose the circle of economics will become more energetic with more ideas in 3–5 years. I am too old to catch up with the trends but I hope you can. Thank you!

Wang Tongsan: Today on the 30th anniversary of the founding of the Chinese Association of Quantitative Economics, every representative has shared his opinion and understanding. In the development history and past achievement, we all mentioned quantitative economics as a discipline with Chinese characteristics. And then we analyzed all the problems that have to be tackled before looking towards the promising future of both the Association of Quantitative Economics and quantitative economics with many good thoughts, suggestions and approaches. In addition, Professor Zhang urged us to pay close attention to future trends after the world financial storm, which is our hope and dream. Chinese Quantitative Economics and the Chinese Association of Quantitative Economics will certainly have a bright future!

The Course of Establishing and Development of the Chinese Association of Quantitative Economics

Zhang Shouyi

Date: May 26, 2009

Why should we study quantitative economics? Why did we establish the Chinese Association of Quantitative Economics? Actually, it has something to do with the current situation of economics in China. After the founding of the new China, economics in China basically relied on introduced research, which fell short in theoretic quantitative analysis. During the construction of China, in the economic field in particular, we were faced with huge and various tasks. Later, we came to realize that to have an effective management, our work should not be limited to just "what" and "why", but to answer the question of "how many", that is a matter of quantitative relations. Therefore, it is a necessary and urgent commission to carry out the quantitative economic research in not only theoretic studies, but also in economic management.

Both the national and international background, then, gave the possibility and necessity for the establishment of the Chinese Association of Quantitative Economics. Inside the country, the Great Leap Forward Movement in 1958 was a disaster for the Chinese people. However on the other hand, it forced many mathematic workers to go outside labs and classrooms and to join in practical management work in factories and villages. That made possible the founding of Systems Research (now Systems Institute).

In 1982, the CAQE held its first annual meeting where 100 students in the workshop played an important role. The second annual meeting was held in Hefei in 1984 and officially announced the change of the name. Since then, annual meetings have been successively hosted in Xi'an Jiaotong University, Jilin University, Shandong Economic University, the Chinese Academy of Social Sciences, Huaqiao University, Chongqing Technology and Business University, Hunan University, Southwest Jiaotong University, Nanjing University of Finance, Zhejiang Gongshang University, Wuhan University of Technology, Ningbo University, and Shenzhen University, each witnessing the growth of the association.

Thirty years of development since the establishment of CAQE has given a new meaning to quantitative economics, which can be comprehended in both a narrow and a broad way. Quantitative economics in a narrow way refers to the research under the guidance of Marxist economics on quantitative manifestation, relations, changes and regularity of changes via mathematical methods and computer science. In a broad sense, quantitative economics covers mathematical analysis, econometrical analysis and modeling analysis. Actually, it can be called a complex subject which is developed based on a basic economic theory and aimed at studying mathematical and quantitative relations via mathematics and computing. In addition, mathematical economics is different from quantitative economics, the former centering on qualitative analysis, theory studies and function relationships among economic phenomena while the latter on quantitative analysis, application research and quantitative relationships among economic phenomena.

A Review of the 30 Years of the Chinese Association of Quantitative Economics

Wang Tongsan

Date: May 26, 2009

Thirty years have passed in a flash but things happened are still vivid in my mind. To commemorate the 30th anniversary of the founding of the Chinese Association of Quantitative Economics (CAQE), we sincerely welcome you to join us on this significant occasion.

I might not remember all about the development of CAQE, so I would like to give a brief review on it. We will have Mr. Zhang Shouyi to tell us more in detail especially about the beginning of the establishment of CAQE.

On March 30, 1979, 18 senior experts including Mr. Wu Jiapei and Mr. Zhang Shouyi founded the Chinese Quantitative Economics Research in Beijing. That day became the birthday of CAQE. Now it is regretful that some of the 18 scholars have left us forever. They are Mr. Wang Hongchang, Mr. Li Bingquan, Mr. Wang Zhao, and Mr. Zhong Qifu. Apart from these, we know that there are among the 18 experts Mr. Qin Wanshun, Mr. Li Yiyuan, Ms. Xue Ying and Mr. Zhang Xuansan. Unfortunately, we have failed to collect information about all 18 people but we are still working on it. It is our hope that one day we can restore the history completely. On this occasion, I suggest that we pay our respect and appreciation to their dedication. Three years after its establishment, that is February 22, 1982, CAQE held its first symposium namely its first annual meeting in Xi'an. Over 150 people attended the conference, including front-line staff in quantitative economics and educators from all over China. Xu Dixin, the then deputy rector of the Chinese Academy of Social Sciences, and researchers Yu Guangyuan and Ma Hongdu gave speeches at the conference. Four topics namely establishing quantitative with Chinese characteristics, better promoting the application of the input–output approach, setting up an econometric model for China, and carrying out economic forecasting were discussed. The successful hosting of this conference was a milestone representing that the Chinese Quantitative Economics Research had become a nation-level academic community.

The second annual meeting was held in Hefei from October 10–16, 1984. More than 130 representatives and 11 non-voting delegates were present. At this meeting, the Chinese Quantitative Economics Research was renamed the Chinese Association of Quantitative Economics. The new name, as the conventioneers believed, would show more clearly the main task of the association which was to host academic activities related to quantitative economics and initiate a new phase in quantitative economics research in China. The CAQE has hosted 16 annual meetings so far. The first one was co-organized by Shaanxi Commission of Science and Technology in 1982 and Shaanxi Academy of Social Sciences in 1982; the second by Anhui Commission of Science and Technology in 1984; the third by Wuhan University and Zhongnan Finance University[1] in 1987; the fourth by the Chinese Academy of Social Sciences in Beijing, 1990; the fifth by Shandong Economic University in Jinan, 1993; the sixth by the School of Commerce at Jilin University and Jilin Socialist Academy in Changchun, 1997; the seventh by the Institute of Quantitative and Technical Economics at the Chinese Academy of Social Sciences in Beijing, 2000; the eighth by Chongqing Technology and Business University in 2001; the ninth by Huaqiao University in Quanzhou, 2002; the tenth by Hunan University in Changsha, 2003; the eleventh by Southwest Jiaotong University in Chengdu, 2004; the twelfth by Nanjing University of Finance in Nanjing, 2005; the thirteenth by Zhejiang Gongshang University in Hangzhou, 2006; the fourteenth by Wuhan University of Technology in Wuhan, 2007; the

---

[1]Zhongnan Finance University was merged into Zhongnan University of Economics and Law together with Zhongnan Law Academy in 2000.

fifteenth by Ningbo University in Ningbo, 2008; and the sixteenth by Shenzhen University in Shenzhen, 2009. Moreover, there is another big occasion in 2009 that is today's commemoration.

The CAQE has made a great number of achievements in the past 30 years. We have 46 collective members, 2036 individual members, and 6 professional committees including econometrics, mathematical economics, input–output analysis, enterprise management and innovation, economic game theory, and quantitative finance. Eleven provincial associations have been set up in Heilongjiang, Jilin, Liaoning, Tianjin, Shandong, Zhejiang, Fujian, Hubei, Jiangxi and Sichuan. Moreover, excellent papers have been collected and published in *Quantitative Economics in the 21st Century* every year since 2000. Up until now, 9 volumes have come out. In addition, the issuing of the Excellent Papers Award and the Outstanding Contribution Award has received positive feedback from society and academia. The former has been issued twice and the latter fourth. The fourth Outstanding Award went to organizers of annual meetings. This time we will take into account individuals. And recently, we are considering setting up the Distinguished Academics Award.

The secretariat opened the official website and membership blog where members can be informed of notifications and activities of CAQE and exchange ideas.

Currently, members of CAQE are dominated by scholars from colleges and universities. For example, Mr. Li Zinai who is a teacher cannot join us today as he has lessons. He feels really sorry about it. So am I, for I think his review on the development of CAQE at universities would be the best. I suggest that the teacher at Jilin University who is going to talk about this topic focus on achievements universities have made so far.

The past 30 years of growth is a crystallization of work by people of all ages. From now on, we shall uphold what the senior members have done for the association, carry on the advantages they have created, and continue to offer the young more opportunities to make the best use of their talents. This year's annual meeting in Shenzhen hosted more than 500 people, most of them being young and active. The rocketing increase of members from 100 to 500 shows the upward development trend of the association. Besides this, we have paid more attention to the academic exchanges between scholars at home and abroad. We have invited several well-known specialists including Nobel Prize winners in Economics to come. This is also one of our achievements.

As for the future of the association, I believe that it relies heavily on the concerted efforts of all members due to the very nature of community. Furthermore I suggest that we focus on the following aspects: Firstly, tightening the information exchange among members; secondly, recruiting more members, in particular collective members; and thirdly, establishing a closer relationship with provincial associations of quantitative economics, which is a key task for the secretariat.

Lastly, on behalf of CAQE and its members, I would like to express my gratitude to the organizer, Beijing Information Science and Technology University. Moreover, I also want to thank Vice President Han Qiushi and Director Ge Xinquan, without whose support it would be impossible for us to gather here today. So let us give them a big hand to show our appreciation.

Part Four

Remembering the Workshop in the Summer Palace*

Liu Hong and Zhang Shouyi (Ed.)

Even the Penglai Island in Summer Palace could not escape from the intense heat of the summer in 1980. Despite this however, more than 100 people gathered there for the seven-week Econometrics Workshop in Summer Palace. Seven economic professors gave lectures, including Lawrence Klein, T.W. Anderson, A. Ando, Gregory Chow, Laurence Liu, Vincent Su and Cheng Hsiao. Veteran scholars like Wu Kejie from Nanjing University and Lin Shaogong from Central College of Engineering were also among the students. Though, this workshop seemed to be a crash course or a literacy class for most of the students. Taking account of its significance, the Institute of Quantitative and Technical Economics at the Chinese Academy of Social Sciences has celebrated its 10th and 20th anniversary respectively in 1990 and 2000. And the academia deems it as influential as "the first class in Huangpu Military Academy".[1] It might be the only workshop that has received such glory and appreciation in the history of economics. After all,

---

*This article is reprinted from the issue on May 29, 2009 of *Economic Observer* with the approval of the author and the newspaper. Zhang Shouyi modified and supplemented some expressions and factual information.

[1]Located on Changzhou Island, Huangpu, Huangpu Military Academy was a new-typed school for the revolutionary military officers. It was set up in 1924 by Sun Yat-Sen with assistance from the Soviet Union and the Communist Party of China, taking "Create the revolutionary army to save the periling China" as its philosophy and 'Love and Good Faith' as its directive. It moved to Nanjing in 1930. The school's major building on the original site was destroyed by the Japanese air bombing in 1938, and the rest were seriously damaged. After the founding of the People's Republic, the original site was put under protection and repaired for several times with government loans.

it represents enthusiasm, initiation and breakthrough, being a cornerstone of quantitative economics. We owe all this to Professor Klein, the initiator of the econometric model and Nobel Prize winner. He took notice of China not only because of the intention of including China into his LINK model, but also because of his experience of being an American communist in the 1930s. Klein was forced to leave America for Oxford during the upsurge in McCarthyism. Until the establishment of diplomatic relations between China and America in 1979, he could lead American economists to China. The first time they came, they handed in economic theses however without any feedback from the Chinese side because Chinese economics had long been left behind. Later, Klein negotiated with Xu Dixin, the then Vice President of the Chinese Academy of Social Sciences (CASS) and Director of the Institute of Economics at CASS, and decided that an econometrics workshop led by Klein should be held in China the next year. Xu Shengwu, the Vice Director of the Institute of Economics took charge and Zhang Shouyi made arrangements for the opening of the class. Zhang Shouyi was first exposed to economic-mathematic method when he was at the Moscow National Institute of Economics, and was allocated to the Economic-Mathematical Method Research Group by Sun Zhifang.

When starting a class, the first thing is to select the site. Since Chinese people care about "good face" very much and want to show friendship to foreigners, Hangzhou, which is considered the most beautiful city in China, was selected first but cancelled due to inconvenient traffic. Later Zhang considered the Summer Palace in Beijing but thought it would be too fantastical to host a workshop in a royal garden. However, unexpectedly, those in charge of the Summer Palace approved his proposal and welcomed such academic activity in light of Chinese reform and opening-up. They offered some rooms on Penglai Island and the site was settled.

The building of the Bureau of Standard Measurement stood just opposite to that of the Institute of Economics. Upon the news of the hosting of the econometrics workshop, they came to complain, "This is supposed to be our business. Why are you meddling with it?"[2] Zhang Shouyi explained

---

[2]This misunderstanding arouse because both "standard measurement" and "econometrics" are translated into the same word "ji liang" in Chinese but with different meanings.

that they were mistaken and what they were in charge of was not what this workshop was centered on. However, they did not buy it and tried to compromise, "Ok. We would not care about your side's being the organizer as long as you agree that our staff should make up more than a half of the preparatory group." But Zhang Shouyi refused the proposal and any one of their staff. Every time Zhang remembered this episode, he would laugh at it, "the Chinese people then had never heard about econometrics. They thought it was the scale to measure weight!"

On June 24, 1980, the workshop had its opening ceremony in the CPPCC Auditorium, after which students were sent to the Summer Palace. As Gregory Chow recalled, "The average age of students was high. Most students were 40 to 50. Some even reached 60. The youngest might be over thirty." I got the roster used then with the help of Zhang Shouyi and found that 30% of 100 students were above 50 and almost 40% were 40 to 50. They were either economic majors with little knowledge of mathematics or the other way around and even economic majors knew little about modern economics. However, professors from America were not aware of this. Klein taught introduction to econometrics including the American economic model, Anderson, probability theory and mathematical statistical analysis, Lawrence Liu, demand analysis, production theory, and Chinese econometric model research, Gregory Chow, econometrics, control theory and vehicle demand function, Cheng Hsiao, econometric method, Ando, application of econometrics, and Vincent Su taught macro-econometric model and forecasting, one professor being in charge of one-week lessons. The trouble was that students barely understood what they were teaching. So to solve the problem, Zhang Shouyi invited Li Chulin from Central College of Engineering to teach mathematics and Li Yining from Peking University to teach economics at night. Zhang remarked, "The lessons of Li Yining were vivid and popular among students." What's more, it was Professor Lin Shaogong at the age of 58 that touched Zhang most. Lin got a doctor's degree in economics from the University of Illinois, used to lecture statistics, economic theory and American economic history and finally returned to China in hope of making contributions to the construction of China. During the workshop, Lin gladly helped translating taped lectures and transcribing them into printed handouts. In this way, the students' learning became more efficient.

The night of July 4 was a night for celebration. A conference was held in an open space in the Summer Palace to celebrate Independence Day. Professor Klein delivered a speech where he remarked, "Nowadays, western countries are suffering from an economic downturn. The Soviet Union and Eastern Europe are having a difficult time as well. But here in China, the economy is rocketing after the reform and opening-up. If it continues, China's future is very promising!" Recalling his words after 29 years, I cannot help exclaiming that how wise and right his words are.

In return of their free lectures, we offered them a trip to any place they wanted. Most of the American professors chose Shanghai or Guangzhou except Lawrence Liu who insisted to visit Dunhuang in Gansu Province. Zhang Shouyi is still amazed by the strong will of Liu until now. The office of foreign affairs at CASS had no choice but to ask Gansu Academy of Social Sciences to look after him. Liu at the age of 35 was the youngest of all the professors, although it was still a tough journey for him. He had to take a train, buses and a carriage on his way and then walked there. Despite all this, he seemed to enjoy his trip very much.

In Laurence Chow's memory, the physical conditions in China were not good, "They often drove me back to the hotel for lunch and did not agree to my request to have lunch with students. One day I finally got the chance but I found that the food and students dining with me were specially arranged." However, Zhang Shouyi is not on the same page about this, "Chow knew little about us and made a fool of himself many times. We were taken good care by the administrative committee of the Summer Palace. Our food was good but Chow thought that we were having rough food every day. One day he suddenly entered the canteen and had lunch with us. After that I asked him about the food. He said 'pretty good'." Though for this story they have different recollections, the effort of all the people who took part in the workshop to get closer and to get to know more about each other are clearly shown.

One day before the morning class, Hsiao asked, "For my lessons I need an article that was published on an American magazine called *The Econometrician*. Where can I find it?" Zhang Shouyi asked him to write down the time information and title of the article and called at once the library of the Institute of Economics. Two hours later, the article was sent to the Summer Palace. Hsiao was deeply moved, "China's modernization would surely be completed ahead of schedule if all people worked as efficiently as you do."

There was no air conditioner in the classroom so they bought some electric fans. Anderson was an academician at the National Academy of Sciences in America and the oldest of the 8 professors. He was very serious about his teaching, chalking the notes on the blackboard carefully. His face became white from wiping away the sweat with hands covered with chalk powder. He asked for an iced beverage but tea was the only drink that could be offered to foreign guests. Zhang Shouyi reported this to Vice Director of the Institute of Economics and got the approval for buying from the administrative office in the Summer Palace a boxful of soda water every day for Anderson, for which the latter was grateful. Later the Office of Foreign Affairs came to know about this and condemned the action; Zhang had to write a letter of self-criticism because of it. In his letter for the tenth anniversary of the workshop in 1990, Professor Hsiao wrote, "I still and will never forget that 10 years before we were working and studying together in the scorching days. Ten years have passed and China has its own promising research path for quantitative economics. I really admire all of you in China for what you have done in spite of hardships and difficult conditions"

Klein borrowed a middle-sized computer that weighed about one ton from IBM. Many students helped to transport it to the Summer Palace. The house where the classes were opened was a two-story building, the upper story being the place where the Empress Dowager Cixi once enjoyed the cool and the lower story the classrooms. They fastened the machine to one end of a rope and manage to drag it up to the second floor, only to find a mismatch in voltage. So Klein had to borrow a one-ton transformer from Japan. People once again managed to bring it back to the Summer Palace. Unfortunately, as it was the first time in life he had seen one, a young man was so curious about the computer that he started it in the middle of the night without approval and mistakenly burned the memory off. The two machines had to be sent back before being used.

The workshop had already started when Mao Yushi at the age of 51 from the Chinese Academy of Railway Sciences heard the news. He went to the Summer Palace, expressing his wish to join in and he was allowed to attend the class. He said, "We were all keen on the study. After lessons in the day, we would have discussion and sometimes make reports. Once

I delivered a report about optimal allocation. " He also told a joke, "People then had not drunk mineral water before. One day a student had some of the water from a bottle left by a professor and then told us, 'It tastes the same!'" In Mao's opinion, the primary significance of this workshop was that it broke through political obstacles and brought western thought to China.

Wang Guangmei, the then director of the Office of Foreign Affairs at CASS, paid a visit to the workshop. Zhang told her, "This is an econometrics workshop, teaching the application of mathematics and computing to economic issues." She was a little surprised, "What a great thing that we've had so many advanced theories and methods!" She listened to one lesson by Professor Hsiao and had a photo with him during the break. Hsiao was very grateful for being able to take picture with the wife of Liu Shaoqi, the then vice chairman of China.

Interesting and funny occasions punctuated the intense study. At times, in order to avoid a fine for a timeout, travelers to Kunming Late would put their boats aside right before the Penglai Island. Some misbehaved young students would hide the boat beneath the water with huge stones and then lifted it up at night so that they could row on the lake. Zhang Shouyi recalled this with pleasure, "I was on the boat once. It was really a scene. The bright moon and cool breeze, the little boat and the tranquility of night, it was wonderful, and different to the daytime."

The workshop came to an end in August. Since the one-hundred-thousand funds have not been used up, Zhang Shouyi invited over 20 journalists from the *People's Daily*, *Guangming Daily*, and the Xinhua News Agency to have a three-day interview. Zhang hoped that via interviews the workshop and, more importantly, quantitative economics would be known to more people. In mid-August, the right upper corner where quotations from Chairman Mao Zedong were placed during the Cultural Revolution was taken up by quantitative economics. "This is an important corner!" Zhang said it with pride. *Guangming Daily* published "Carry out Research and Application on Quantitative Economics", pointing out that "The past mistakes in decision making have something to do with the neglect of quantitative research and analysis and lack of proof in decision making." It also published the speech by Ma Hong that was titled "Quantitative Economics: A Useful Subject for Socialist Construction". Thus the cause of championing quantitative economics had started. Mr. Lin Shaogong and Li Chulin began to teach Econometrics and Mathematical economics at the Central College of Engineering. Many colleges and universities started to carry our teaching and research activities. Research centers were established and books and magazines about quantitative economics gradually came out in increasing numbers.

In September 2009, the international symposium themed Quantitative Economics and its Application in Twenty Years was held in Beijing. Seven of the eight professors at the 1980 workshop came, so did many of the students. They were all growing older but their careers since 1980 were rocketing. Academic articles submitted to the conference were published in the book *Frontier Quantitative Economics*.

"How about hosting a 30th anniversary next year, the year 2010?" I suggested to Zhang Shouyi, who became the President of CAQE and an honorary member of CASS, and who is now already in retirement. "That may depend on whether Klein can make it," he replied. Alas, Professor Klein though always look younger than his years will be 90.

Postscript

Thanks to the reform and opening-up policy, the Chinese Association of Quantitative Economics has greeted its 30th birthday with ever-growing vitality. It had 18 specialists at its founding and 30 years later it now has more than 500. I dare to say that it just like a sapling has grown into a huge tree with flowers and fruits.

We should first express our gratitude to Beijing Information Science and Technology University with whom the publication of the *festschrift* *30th Anniversary of Chinese Quantitative Economics* would be impossible. In 2008, Professor Ge Xinquan from BISTU proposed the hosting of a 30th anniversary commemoration. On May 26, 2009, his proposal was realized under the organization of BISTU. More than 40 specialists and scholars attended the conference, including Han Qiushi, the Vice President of BISTU, Ge Xinquan, the Director of the School of Economics and Management, Zhang Shouyi, the Honorary President of the Chinese Association of Quantitative Economics, and Wang Tongsan, the President of CAQE. The meeting decided to publish a *festschrift* to review the 30-year development of the association and to act as a reference for future development.

Articles in the *festschrift* were written by members of CAQE, many of whom have witnessed the growth of the community. Their writings cover not only theories of quantitative economic, but also the application of quantitative economics to real economic practice, summaries and reviews on the development history of CAQE, and their reflections based on specific historical periods.

The editing work of this *festschrift* has received great support from both initiators of Chinese quantitative economics and promising young scholars. Wang Guocheng from the Committee of Game Theory and Experimental

Economics, and Liu Qiyun and Xia Ming from the Input–Output Committee contributed information and materials about their branches to the editing of *The Record of Major Events of Chinese Association of Quantitative Economics for 30 years* (*1979–2009*). Moreover, I would also thank Hao Chunhe who, suffering from eye disease, hand-wrote the section about the Enterprise Committee without seeking any payment.

Lastly, suggestions for revisions of this book are sincerely welcomed.

The Editors
November 2009